PULSE OF THE PLANET No.1:

On A-Bombs, Polar Motion, Cloudbusting, Droughts, and FDA/"Skeptic Club" Slanders of Wilhelm Reich

The occasional research report and journal of the
Orgone Biophysical Research Laboratory, Inc.

Editor: James DeMeo, Ph.D.
Production Assistant: Theirrie Cook, B.A.

Volume No. 1
Spring 1989

CONTENTS: **Page**

PULSE OF THE PLANET No.1:

ISBN: 978-0-9891390-4-5   ISSN: 1041-6773

The occasional research report and journal of the
Orgone Biophysical Research Laboratory, Inc.

Distributed by
Natural Energy Works / Ingram/ Lightning Source

Send all communications to:

PULSE OF THE PLANET
Care of James DeMeo, PhD, Editor and Director
Orgone Biophysical Research Laboratory, Inc.
Ashland, Oregon, USA
demeo@orgonelab.org

Graphics courtesy of *Art For Environment's Sake*, published by the Environmental Task Force.

150607

Editor's Page

Welcome to the first issue of *Pulse of the Planet,* the research journal of the Orgone Biophysical Research Laboratory. The idea for this journal developed over the last several years, as the social/ environmental crisis swelled to global proportions, and as conventional solutions offered to the crisis were clearly demonstrated to be insufficient. The *Pulse* takes the position that Dr. Wilhelm Reich's sex-economic findings, and his discovery of the biological and atmospheric life energy, are crucial for understanding the basic nature of these massive environmental and social problems. Most of his findings have been independently corroborated and strengthened in recent decades, though little of the newer research has broken into the mechanistic science research journals, or into the popular media. Interest in his research is nevertheless widespread, and growing. In large measure, this is due to the republication of Reich's books, and to the publication of additional corroborating research materials over the last several decades, mainly in the *Journal of Orgonomy.* Additional journals and publications devoted to orgonomy, the science of orgone (life) energy functions in nature, have appeared in recent years. Educational workshops and college courses focused upon Reich's life and work are also now offered around the nation. In Europe, Reich is even better known than in the USA; there, public lectures on his works draw hundreds of people, and orgonomic research proceeds with an institutional tolerance and support that has yet to fully materialize in the USA. All of this has fueled interest in the man and his works. However, *Pulse of the Planet* will fill a different role and function from previous publications.

The *Pulse* will largely focus upon environmental problems, and their underlying social-emotional roots, as they are developing. Our research reports, on nuclear testing, weather extremes, earthquakes, desert spreading, pollution crises, and other unusual phenomena, should be of interest to both layperson and researcher. This focus upon the environmental and social background is vitally necessary, given the impending global disaster that is brewing, from centuries of unrestrained population growth (a product of sex-repression), and the damaging environmental practices of an emotionally armored humankind. Reich foretold of this developing crisis, which is projected to worsen until around the year 2020, when lack of food, water, and natural resources will likely force a precipitous halt and decline in global population. Some argue that this is already starting to occur in the overpopulated and impoverished parts of Africa, Asia, and Latin America. Global reports on these subjects will be carried in the *Pulse.* Unlike other environmental journals, however, the *Pulse* will focus upon the available tools and solutions for these problems that orgonomy and other new research developments have to offer. These new discoveries can make a considerable difference for life on Earth, assuming they are rationally applied.

The *Pulse* will also be a resource and networking guide for persons interested in life energy functions, and environmental issues. Research papers from workers and associates of the Laboratory will be published, as will other important matters that come to our attention It is hoped that the *Pulse* will open lines of communication between those in the field of orgonomy, with those in other natural scientific branches. There is a growing need for accurate and timely information on these subjects. *Pulse of the Planet* will help to build some bridges, and to fill this need.

The Editor
March 1989

Editor's Postscript to the 2015 Reprint Edition

Some 26 years have passed since the first numbered edition of *Pulse of the Planet* was released. The original edition was issued in a limited run of 500 copies, which were sold or given away quickly, after which we never had the funds, nor time and energy, to undertake printing up additional copies. It was only made available thereafter as a photocopy. The internet matured over the years and our small non-profit foundation began using it, as well as new print on demand methods (POD) to circulate our publications. It was then decided to release a reprint edition of *Pulse* by POD, for historical purposes.

Looking back at this early work, the pioneering subject material is still important for the modern world, in spite of major social and international changes. Atom bomb testing has ended. The Cold War also ended with collapse of the Soviet Union. Internet came available as a means for lesser-heard voices to level the playing field in the struggle over ideas. The 911 terror attacks happened, initiating major international developments which, by 2015, have placed the world on the brink of yet another global war. Wilhelm Reich's natural scientific orgonomy has continually been tested out scientifically over these 26 years, with many positive replications, but the "skeptics" war of slander and misinformation continues as loudly as ever. Nevertheless, the walls of irrational opposition are crumbling. Regarding my own work, my Orgone Biophysical Research Lab (OBRL) found a permanent home for a research laboratory and educational center in the pristine forested mountains of southern Oregon, near to Ashland. And that has provided many new opportunities and scientific developments.

I also was forced to ask, before undertaking this reprint edition, how much of anything needs correcting? The answer is, surprisingly little, though surely many aspects of the subject material could use an updating. However, aside from this 2015 "Postscript" page, and some strike-throughs of old dead addresses and such, this reprint edition is unchanged from the original. But the following notes will be valuable to know, as helpful stepping-stones for the interested reader.

Regarding the original Editor's Page, written in 1989, I surely was at that time persuaded by some of the more spectacular claims of the environmental movement, notably that petroleum resources were depleted and that would act as a brake upon human development. Instead, there is more petroleum being pumped and refined today than ever before.

Also, I held the view that population growth would follow a Malthusian pattern, fueling a dramatic depletion of food and other resources. While human populations as a whole do continue growing, the growth is not uniformly distributed with some regions still growing fast, and others with declining numbers. The steady growth in contraceptive usage globally has applied the brake, reducing growth rates to more manageable levels. Also, pressure on high population regions has been relieved by emigration, notably from Latin America, Africa and Islamic regions, abandoning impoverished homelands and moving into Europe and North America for better economic opportunity. This has benefited those immigrants tremendously, but has generally been a deficit to the host nations, in how uneducated and monolingual arrivals, with few job skills and demanding social welfare, often boasting an attitude of superiority over their host nations, generally bring more economic loss and crime or terrorism than benefits. And while many environmental problems have been mitigated, the critical factor of depletion of world fisheries continues, being harvested by industrial machinery to feed an increasingly hungry world. So while the global population catastrophe continues, it has transformed into a different form, shape and nature than originally anticipated.

Regarding the paper "Cloudbusting: A New Approach to Drought", this was written at a time when global deserts were expanding, and both droughts and heatwaves were frequent and intensive. We had good reason to believe a-bomb testing was exacerbating this process, creating flash-point heatwaves via Reich's *oranur effect*. Consequently, *Pulse* included new reports on unusual atmospheric and biological reactions to a-bomb testing. This and subsequent issues of *Pulse* carry a lot of important information on that subject. However it must also be noted that starting around the year 1998, which was a peak El Niño year, global temperatures stabilized and Antarctic sea ice has reversed its decline, increasing by all available measures. The year 2000 also saw a reversal of desert expansion in North Africa, as discussed in this first article. There is now evidence of a return to slightly wetter conditions in some parts of the large *Saharasia Desert Belt*. Some of those changes may have been stimulated by the 12 separate cloudbusting expeditions organized by the author into Israel and Africa in the 1990s.

Many new cloudbusting operations were launched through OBRL since this original issue of

Pulse. There was a systematic field trial in Arizona, and major drought-relief efforts in both Israel and Namibia, Africa. A multi-year project was initiated from 1994 through 1999 in the nation of Eritrea, in the Horn of Africa. All of these efforts ended serious droughts with significant increases in rainfall, and were well-documented efforts, about which numerous articles and book chapters have already been published. The Saharasian Desert Belt and those newer cloudbusting expeditions are also discussed in subsequent issues of *Pulse*, and in other published articles available from the weblinks to follow.

Regarding the paper by Kato and Matsumae, this item created a stir among environmental and anti-nuclear organizations internationally, who were alerted to the findings. Similar articles on anomalous geophysical and atmospheric reactions to *underground* a-bomb tests and nuclear power plant accidents were published in later issues of *Pulse*, all of which were circulated widely, suggesting they may have played a role in ending of atomic bomb testing. Those articles were so obscure that, prior to our publishing them, few even knew about their existence. I nevertheless consider the study of Kato and Matusmae as a preliminary effort, requiring corroboration by retrospective analysis.

Regarding the article on Martin Gardner and the skeptic club attacks against Reich and myself, I wish I could say they had ended. Today there are even worse slanderous things being circulated on internet. But the truth has also been equally well-circulated, discrediting those organized bully-boys in the "skeptics" groups. So it is clearly a better situation than in the past, when no mainstream newspaper or publishing house would dare to seriously publish anything which presented the facts about Wilhelm Reich's work. Martin Gardner also died in 2010, never apologizing for his slanders against Reich, or for his role in stimulating the fraudulent FDA "investigation" of Reich, and the burning of his books and research journals.

Regarding the "Postscript to the FDA" article, this was a follow-up to the excellent studies undertaken earlier by R. Blasband, C. Baker, and J. Greenfield, all of which exposed the FDA's bias and unscientific approach to their "investigation" of Reich's *orgone energy accumulator*. In particular, I addressed some new revelations about one of Reich's hidden critics of those days, an MIT physicist, and deconstructed his claims to have replicated Reich's critical To-T experiment, among others. Today, there is even better, very robust and definitive experiments which prove the reality of the To-T effect, or thermal anomaly inside the orgone accumulator.

PDF downloads of those research papers are available at my academia.edu website, which is given below.

Regarding the various "Climate Features" maps, these were developed from an on-going effort at OBRL, towards a better understanding of the fluctuating nature of climate and related cosmic, atmospheric and geophysical phenomena. Global patterns were monitored closely at my laboratory, and put to use in various contexts, as in field work with the cloudbuster in overseas expeditions during the 1990s. However, organizing such data for public presentation became too time-consuming for our limited resources, and so was ended after a few years. Today there are scientific institutions and private efforts making similar information available through the mainstream media and internet.

I can also mention several new books were recently published, exposing the people behind Reich's persecution and death: James Martin's *Wilhelm Reich and the Cold War*, and my own *In Defense of Wilhelm Reich*. Both are available from booksellers.

Orgonomy as a natural science has emerged from the shadows of the "skeptic" slanders and FDA book-burning. Today there are mainstream publishers and science journals allowing open discussion of the issues and facts developed in Reich's original works. One can actually speak and write about the orgone energy accumulator in scientific forums, and publish research articles on Reich's findings in more open-minded peer-reviewed journals.

The various issues of *Pulse of the Planet* played a small but important part in these social transitions, and so we are happy to republish them.

James DeMeo, PhD
Editor, Pulse of the Planet
Director, Orgone Biophysical Research Lab
Ashland, Oregon, USA
May 2015

Updated Contact Information:

* James DeMeo's Research Publications:
    http://orgonelab.academia.edu/JamesDeMeo
* Orgone Biophysical Research Lab:
    http://www.orgonelab.org
* Saharasia web page: http://www.saharasia.org
* Natural Energy Works, books:
    http://www.naturalenergyworks.net

Cloudbusting:
A New Approach to Drought*

James DeMeo, Ph.D.**

[* An earlier version of this article first appeared in the August, 1988 edition of *Acres USA* (PO Box 9547, Kansas City, MO 64133), with a subsequent reprinting in *Wildfire* magazine (PO Box 9167, Spokane, WA 99209). Both are excellent publications, and the reader is encouraged to subscribe to them.]

Overview of the Environmental Situation:

The year 1987 hold the dubious distinction for being the hottest on record, worldwide. The previous hottest year of record was 1981, and six of the hottest years of record within the last century have occurred within the 1980s. These facts have awakened even the more conservative climatologists to a harsh reality: Our atmosphere is exhibiting signs of dysfunction, and not merely "natural variations" or "anomalous" extremes. One fact has yet to become more widely appreciated. Global warming is not proceeding in a slow, gradual or evenly-distributed manner. Some regions of Earth are periodically heating up very fast, in local surges or episodes that we call *droughts* or *heat waves.*

In the last few years, for instance, we have seen major droughts of historical, or record-breaking proportions: in the American Midwest and West, in southern Europe, India, China, and in the tropical rainforests. Massive forest fires of never-before-seen proportions have accompanied these droughts in many cases. Other signs of widespread atmospheric disturbance, or pathology in the biosphere, exist: several gargantuan icebergs have broken away from the Antarctic ice shelf, possibly as a consequence of global warming and an observed slight increase in the sea level. One of these icebergs is the size of the state of Rhode Island, while another, of Connecticut; both have run "aground" on the continental shelf off-shore of Antarctica, while a third behemoth iceberg, the size of Hong Kong, was recently spotted adrift in the open oceans. These events take place at a time when mountain glaciers all around the globe are slowly melting away. The height of Atlantic Ocean sea waves have recently exhibited a measured increase, suggesting more turbu-

lent atmospheric conditions. Meanwhile, Caribbean coral reefs and populations of Atlantic and Pacific sea mammals are dying off, from infectious diseases related to a weakened immune system.

In the atmosphere , we see deterioration in many areas. Aside from drought, levels of pollutants, and windborne dusts from drought parched areas, are increasing the turbidity and stagnation in the lower atmosphere. In the upper atmospheric, protective ozone is deteriorating at a shocking pace, and may soon yield sunlight of a more burning, even killing character. The rainfall in many areas of the world has already become partly toxic in character, killing to both plant and animal.

Added to the above is the distressing fact that some 60,000 square kilometers of grassland or scrubland vanish each year, being replaced by barren desert lands. These include marginal, steeply sloped, or semi-arid drylands that have been subjected to salinity from irrigation, overgrazing, or unsuitable agricultural techniques. Massive deforestation and soil erosion, with a consequent disturbance of atmospheric pulsation and rains, has fueled or accompanied this desertification process, which has been exacerbated by a continuing human population explosion, also of an historically unprecedented magnitude.

The bottom line of this whole distressing process is simply, that our beautiful and unique, wet, blue planet is slowly being converted into a barren wasteland, a ball of dust. In another publication (*Journal of Orgonomy*, Spring 1989) I have shown that the basic form and expression of this process is best seen in the context of continuing desert spreading and drought. In 200 years, if current trends continue, Earth might look more like the planet Mars, a lifeless ball covered with sand dunes and barren wasteland. Indeed, if a science fiction book were to be written about some hypothetical planet where these *preventable* conditions were allowed to persist, we would be forced to conclude that its inhabitants were totally insane. And yet, every fact given above is well documented.

History also supports the trend. The largest and harshest single broad desert region on Earth, what I call *Saharasia* (which stretches across North Africa, the Near

** Director of Research, Orgone Biophysical Research Lab, PO Box 1395, El Cerrito, CA 94530 USA

> **"... our beautiful and unique, wet, blue planet is slowly being converted into a barren wasteland, a ball of dust."**

East, and into Central Asia), was — prior to c.4000 BC — a wet and lush semiforested grassland, with huge lakes and streams that supported a myriad of large and small animals. The sun-bleached bones of those animals today litter the barren desert wastelands, their human populations having long since been driven to wetter borderland regions. Other desert regions have likewise exhibited similar long-term trends towards increasing aridity. The modern-day rates of deforestation and desert expansion are, however, unprecedented. While we may wonder about the causes of ancient desertification, there is little doubt about the effects of human actions in the modern desert formation process.

**New Discoveries Regarding
Basic Atmospheric Functioning:**

There are many reasons to be hopeful, however. Aside from the many very important and necessary proposals being put forward by different environmental groups, there presently exists a relatively new and effective means for eliminating the atmospheric components of drought and desert-related processes. This new means is based upon the discovery, made by Dr. Wilhelm Reich in the 1950s, that a pulsatory, water-attracting energy is at work within both living organisms and the atmosphere. Reich called this new energy the *orgone*, and he developed a number of methods to objectively demonstrate its existence, and identify its properties. His experiments indicate that the orgone is similar to what had previously been suspected to exist as a *vital force* or *aether*. The experimental investigations of many contemporary scientists, including my own, have repeatedly confirmed and verified Reich's claims.

Atmospheric orgone energy, or life energy, for example, is a fundamental factor in determining the regularity and quantity of rainfall. Where the orgone energy maintains a state of lively pulsation, cloud cover and rainfall follows a cycle of regularity, with alternating periods of rain-dry-rain-dry, even if only seasonal in nature. However, where the orgone becomes stagnant and immobile, losing its pulsatory qualities, clouds fail to form or grow to significant size, and rains decline or cease altogether. Under energetically stagnant conditions, such as during drought or massive forest death, the orgone loses its normal blue-green color, and is replaced by a steel-grey or brownish hazy quality, which Reich called *DOR*, short for deadly orgone. The DOR can be seen in droughty or desert regions as a thick haze which diminishes visibility, imparting a darkish coloration to even small, well-illuminated clouds, and a burning, parching

quality to the sun and wind. While classical climatology and meteorology have little to say about these observable phenomena, and cannot offer any real help to life in drought regions, Reich's energetic formulations have led to a breakthrough.

Prior to his death in 1957, Reich invented an apparatus called the cloudbuster, which — when properly used by a skilled operator — can draw off or neutralize the stagnant atmospheric energy (DOR), and restore atmospheric pulsation and rainfall. A properly executed cloudbusting operation, even when performed under drought conditions, is followed by a visible and sensible excitation of the atmosphere. Atmospheric haze decreases, and visibility measurably increases afterward, while clouds develop and grow, or change in form, from the tattered, brownish drought-types (cumulus fractus) to the billowy, cauliflower form necessary for rains (cumulonimbus). These changes usually begin within a few minutes to hours following the onset of cloudbuster operations, and propagate outward to affect regions from the size of a single county, to that of several states in area.

Reich made a number of practical demonstrations of the cloudbuster in the 1950s, successfully terminating droughts across the eastern USA. He also conducted cloudbuster operations in the Tucson, Arizona region, demonstrating that the desert itself could be greened through effective and skilled application of the technique. There is no question that he would have continued with this research, and would likely have made a major demonstration of the desert-greening effect, had he not been pushed to an early death by various hostile, anti-scientific critics.

In the late 1970s, when I was a graduate student and Instructor at the University of Kansas, I systematically tested the cloudbuster's effects upon Kansas weather. The results of those experiments confirmed in many ways that the cloudbuster had definite atmospheric effects. Experimental cloudbusting research since that time, in droughts affecting Kansas, Illinois, Florida, Georgia, the Carolinas, and the state of Washington, has repeatedly confirmed the effectiveness of the cloudbuster in eliminating atmospheric stagnation, and in bringing about a restoration of atmospheric pulsation and regular rainfall. To date, I have participated in, or directed more than 30 separate cloudbusting operations. Approximately 80% of these have been followed by increased cloud cover and rains — often of a most dramatic nature — within 48 hours after onset of cloudbuster operations. Some 16 of these operations proceeded under mild to severe drought conditions, wherein no significant rains were anticipated for weeks or months to come. Even so, the 80% success rate was also observed for drought conditions, and rains

> **"While classical climatology ... cannot offer any real help to life in drought regions, Reich's energetic formulations have led to a breakthrough."**

generally *persisted* following these operations.

The most striking example of these effects was observed during the 1986 southeastern USA drought, which ended following a series of early August cloudbusting operations in Georgia and South Carolina. In that case, drought conditions ended with a long, soaking rain which covered the entire Southeast, lasting nearly two weeks, after which regular rainy and dry periods alternated for months. The weather forecasters and climatologists, who generally ignore the published research data and notifications on the subject of cloudbusting, were completely surprised by the early August ending of that drought, which was forecast to last until October. In another recent experiment, a cloudbuster was tested in the harsh desert regions near Yuma, Arizona. A significant pulse of moisture moved into the area from the Pacific Ocean shortly after that experiment, bringing rains across the desert Southwest, and into the droughty Midwest as well. This latter experiment confirmed in a very powerful way the desert-greening possibilities first noted by Reich in the 1950s.

Drought Abatement and Desert Greening Projects:

Presently, two major projects are under development, for bringing cloudbusting techniques to regions of environmental need in a safe and responsible manner. These projects were developed by the Orgone Biophysical Research Laboratory, in consultation and coordination with other responsible individuals and organizations (notably, the American College of Orgonomy) with interests in Reich's works, and the continuing environmental crisis. The two programs are:

1. A *Drought Abatement Outreach Program*, established and available to service any North American, or foreign region suffering from drought conditions. Mobile and transportable cloudbusters, with trained and skilled operators, can be transported to drought regions to benefit those areas. Proven and safe cloudbusting techniques can be used to restore atmospheric pulsation and rains.

2. A *Desert Greening Project*, currently in the developmental stages. A preliminary series of experimental tests with the cloudbuster in the deserts of the American Southwest will be undertaken, to provide a better data base for the effects of cloudbusting in the drylands. A permanent facility in the desert lands is also being sought for establishment of a base of operations to better demonstrate the long-term effects of desert cloudbusting.

These two projects will hopefully provide a means

for dealing with the atmospheric expressions of the current environmental crisis. It must be emphasized, however, that without a firm and uncompromising commitment to basic environmental principles (recycling, resource conservation, reforestation, species habitat protection, pollution controls, renewable energy systems, ecological agriculture, family planning), cloudbusting will do little more than to temporarily delay an otherwise inevitable global catastrophe. It is not a panacea for a society that might wish to only remedy the more overt symptoms of human psychopathology and greed (armoring). The much larger, and more difficult questions of human attitudes towards nature, and natural biological processes, remains, as always, the central dilemma.

Individuals or organizations with the interest and resources to sponsor either the *Drought Abatement* or *Desert Greening* programs, or with questions regarding the possible use of the cloudbusting technology in their area, should contact the author.

Selected Citations on Cloudbusting:

1. Blasband, R. (1970): "Orgonomic Functionalism in Problems of Atmoshperic Circulation, Part II, On Drought", and "...Part III, On Desert", *J. Orgonomy*, 4(1):62-78, and 4(2):167-182, 1970.

2. Blasband, R.A. (1972b): "OROP Hurricane Doria", *J. Orgonomy*, 6(1):80-83.

3. DeMeo, J. (1979): *Preliminary Analysis of Changes in Kansas Weather Coincidental to Experimental Operations with a Reich Cloudbuster*, Geography-Meteorology Department, University of Kansas, Thesis, 1979(xerox available from Nat. Energy Works, PO Box 1395, El Cerrito, CA 94530).

4. DeMeo,J. (1985): "Field Experiments with the Reich Cloudbuster: 1977-1983", *J. Orgonomy*, 19(1):57-59.

5. DeMeo, J. & Morris, R. 91987b): "Preliminary Report on a Cloudbusting Experiment in the Southeastern Drought Region, August 1986", *Southeastern Drought Symposium Proceedings*, March 4-5, 1987, Columbia, S.C., South Carolina State Climatology Office Publication G-30, pp. 80-87.

6. DeMeo, J. (1989): "Desert Spreading and Drought; Environmental Crisis", *J. of Orgonomy*, 23(1), in press.

7. Reich, W. (1952): "DOR Removal and Cloudbusting", *Orgone Energy Bulletin*, IV(4):171-182.

8. Reich, W. & Moise, W. (1955): "OROP Hurricane Edna", *Cosmic Orgone Engineering*, VII(1-2):84-92.

Editor's Note:

The following paper by Yoshio Kato, with a preface by Shigeyoshi Matsumae, was informally circulated in 1976, but apparently has never before been published. While the study demonstrates only a weak correlation between underground nuclear bomb testing and changes in geophysical and atmospheric phenomena, there is enough here, in the Editor's view, to warrant public discussion and further investigation. The graphed data suggests that the planetary response to underground nuclear testing is greater for larger, more powerful nuclear explosions, and also greater at those times when a number of nuclear bombs are exploded sequentially, one after another. The data for late April and late October are more convincing on these points, than the data for other periods encompassed in the study.

The postulated effects discussed here make no sense whatsoever from the viewpoint of conventional physics, which rests upon the assumption of "empty space". However, they make very good sense, and can be explained, on the basis of an energetic disturbance propagated through a mass-free (but mass-affecting) energy continuum. Reich's orgone energy fulfills these characteristics in a rather complete manner; the orgone is additionally known to be greatly agitated and excited by nuclear radiation, creating an unshieldable energetic disturbance known as the oranur effect, which can propagate over great distances. Unlike ionizing nuclear radiations, oranur effects from underground nuclear bomb tests would not be restrained by overlaying strata of rock and earth, and have been cited as possible causes of widespread atmospheric stagnation and/or tornadoes even before the Kato paper came to the light of day. More will be said on this subject in future issues of the Pulse.

Following the Editor's 1987 discovery of Kato's unpublished paper, copies were sent to various research journals, scientists, and environmental groups in the USA and Europe. A subsequent notice of his findings appeared in the Journal of Orgonomy, Acres, USA, and the Earth First newsletter. This apparent first circulation and discussion of Kato's findings in the USA took place only a few weeks before a large Canadian earthquake, which was shortly preceded by an American underground nuclear bomb test. Two large devastating earthquakes in the central Soviet Union also occurred around that time, and both of them were also shortly preceded by Soviet underground nuclear bomb tests. These observations underscore the need for new research into the question of long-distance, delayed atmospheric and geophysical responses to underground nuclear bomb tests. The stakes are too high to casually dismiss even a "weak correlation".

James DeMeo, PhD
March 1989

Recent Abnormal Phenomena on Earth and Atomic Power Tests

Yoshio Kato *

With a preface by Shigeyoshi Matsumae **

Preface:

Recently, there have been numerous reports of abnormal meteorological phenomena and severe earthquakes occurring so frequently in the whole areas of the Earth, and also the abnormal polar motion of the Earth.

To elucidate this subject, I entrusted Professor Yoshio Kato, Professor in charge of the Aerospace Department, and Director of the Industrial Science Research Institute of our university, with preparations of research material.

Dr. Kato has performed this task exquisitely and has found that the aforementioned abnormal meteorological phenomena, earthquakes, and fluctuations of the Earth's axis are very much related to the atmospheric and underground testing of nuclear devices.

There are, of course, many primary factors controlling and affecting these phenomena. However, the evidence is overwhelming that the recent abnormal phenomena, never experienced before in recorded history, have been caused by nuclear testing. Therefore the nuclear tests being conducted by numerous countries, in competition with one another, must be stopped.

We, mankind, must place the protection of the Earth above and before the protection of our individual countries. This is the common order on which all of mankind must now reflect. If the conscience of science fails in this mission, the world may perish.

At the earliest possible date, all nations of the world should declare an immediate ban on nuclear testing.

Hereinafter the details of Dr. Kato's study are described.

September 30, 1976
Shigeyoshi Matsumae

* Head of the Department of Aerospace Science, Tokai University, Japan.
** President of Tokai University, Japan.

While it must be presumed that everyone feels a sense of crisis or fear over the repeated and almost unlimited testing of nuclear devices, it must also be presumed that most believe that this frequent release of enormous amounts of energy has affects on the nature of the Earth. The remarkable fact is, as obviously proven by our report, that nuclear testing does indeed have a serious affect on the physical structure and environment of the Earth.

Our finding is that the abnormal atmospheric phenomena and recent frequent large earthquakes are not only related to nuclear testing but that there is a direct cause-and-effect relationship.

To begin with, nuclear testing has caused the temperature of the Earth's exosphere to rise abnormally by from 100 to 150 degrees absolute temperature. It is obvious that this abnormal temperature rise affects the atmospheric phenomena of the Earth.

Also, it has been found that nuclear testing is the cause of abnormal polar motion of the Earth.

Changes in Upper Atmospheric Temperature

These discoveries have been made possible largely by Tokyo University Aerospace Institute's fifth man-made satellite TAIYO (TAIYO means SUN), which was launched on February 24, 1975, and has been sending data which can be considered solely the result of Japan's technological efforts. The TAIYO is in an elliptical orbit — 255 kilometers at perigee and 3,135 kilometers at apogee above the Earth — at a 31 degree angle of inclination to the equator, and takes two hours to circle the Earth. Compared with recent foreign satellites, it travels much closer to the Earth. Therefore data from the atmosphere not previously available from recent foreign satellites has been obtained.

Data sent from TAIYO does not include the actual temperature of the atmosphere. This is computer calculated by a complicated formula. Basically, decay of the orbital period caused by atmospheric resistance at approximately 250 kilometers is obtained and variation of the atmospheric density is derived. Since atmospheric density is mutually related to temperature — as temperature increases, density decreases and vice versa — the atmospheric temperature can be calculated. For this the internationally used atmospheric model may be used. The result is further applied to determine the exospheric temperature at a point approximately 1000 kilometers away from the Earth. In this manner, the data in Figure 1.1 (since March 1975) for the exospheric temperature was obtained.

Great fluctuation is clearly present in this graph. To obtain a stable temperature table, the elements which cause the temperature to change, such as difference by night and day, effects of solar activity, etc., must be subtracted from the graph.

Charged particles from the Sun, commonly called the solar wind, change the activity of the Earth's magnetic field and are indicated in units of Kp. Data on these charged particles obtained from over seventy countries is collected at the International Data Center in Colorado, USA. [Now called the Space Environment Services Center] The variation in solar wind charged particles is also graphed (Figure 1.2).

Figure 1: Through computations based on data received from the man-made satelite TAIYO on the atmospheric temperature at aproximately 250 kilometers, Chart 1 was produced. If data on charged particles (Chart 2) and ultraviolet rays (Chart 3) from the Sun are subtracted, the temperature change wave should be reduced to near level. However, the wave does not disappear, but remains as shown in Chart 4. When this is compared with the periods when nuclear tests were conducted, the mutual relation between them is obvious. Particularly in October, when a series of tests were conducted, a radiical temperature rise is evident.

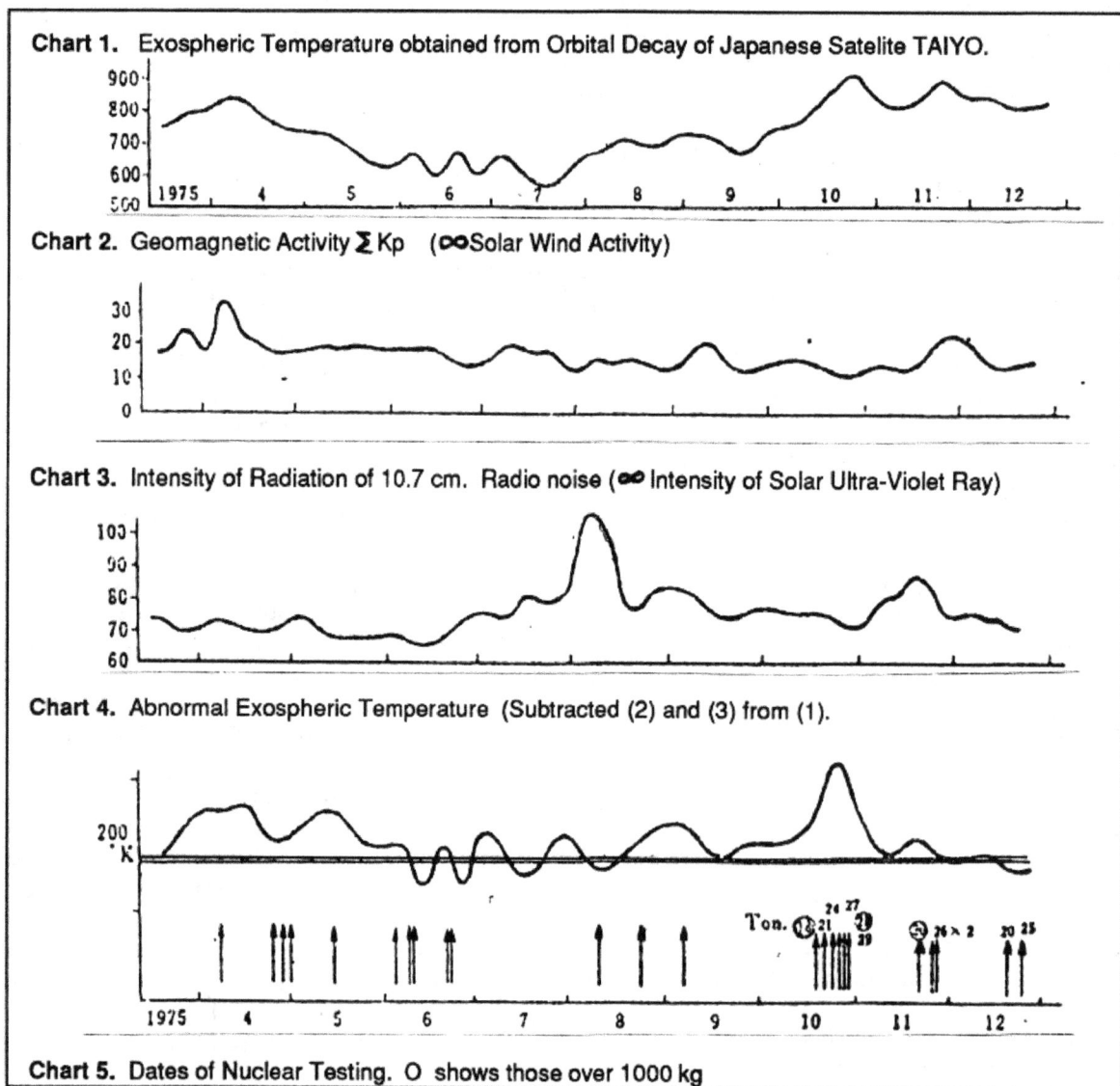

Chart 1. Exospheric Temperature obtained from Orbital Decay of Japanese Satelite TAIYO.

Chart 2. Geomagnetic Activity Σ Kp   (∞Solar Wind Activity)

Chart 3. Intensity of Radiation of 10.7 cm. Radio noise (∞ Intensity of Solar Ultra-Violet Ray)

Chart 4. Abnormal Exospheric Temperature (Subtracted (2) and (3) from (1).

Chart 5. Dates of Nuclear Testing. O shows those over 1000 kg

Data on solar radio noise which is proportional to the intensity of ultraviolet rays, obtained by each of the reporting countries, is collected in Colorado in the same manner. Intensity of energy in the 10.7 centimeter wavelength radiation, since March 1975, are arranged in Figure 1.3.

Now, if the solar activity factors which are considered to cause the temperature to change, that is the solar charged particles (Figure 1.2) and ultraviolet rays (Figure 1.3), are subtracted from the exospheric temperature graph (Figure 1.1), the wave should theoretically be reduced to zero. That is to say that the temperature line should become almost flat. In other words, the temperature change should be equal to the equivalent changes in the factors that cause the temperature change. Therefore, if the later are subtracted from the former, the large wave on the graph should disappear. Quite on the contrary, in actual calculation the wave of the temperature difference remains as shown in Figure 1.4. Therefore, it must be concluded that the wave showing high temperature is not caused by solar activity, but rather by a completely different factor which has never before been considered. When considered from the conventional theory, the result is just not logical.

Then, after the data on nuclear tests (Figure 1.5), obtained from the Foreign Ministry Information Department — data which had originally been considered as unrelated to the matter in question — was plotted, it was found that the graph of abnormal temperature (Figure 1.4) and the graph of nuclear tests (Figure 1.5) almost completely overlap. In other words, the exospheric temperatures rose abnormally immediately after a nuclear test was conducted.

For example, it was found that the temperature rose by seventy to eighty degrees absolute temperature after a nuclear test in the Soviet Union which was observed and reported by Upsala on the 23rd of August last year [1975]. Similarly, continuous and drastic temperature rise was observed at the time of intensive nuclear testing, totalling six separate tests, between October 18 and 29, 1975. It is regrettable that the daily change could not be recorded because data from TAIYO is received once a week only.

In considering the relation between exospheric temperature and nuclear testing, the temperature abnormality was confirmed first, then the data on nuclear testing was plotted.

Changes in Polar Motion and Earthquakes

Besides the above study, I examined the polar movements and the affects of nuclear testing. Although the Earth turns on its axis, the pole (North Pole) is not stationary, but moves in a circle of a few meters. This was discovered by Chandler at the beginning of this century. The polar movement, which makes a revolution in 430 days, is called the 'Chandler Cycle'. Presently, every country at north latitude 39 degrees has an observatory at that latitude for continuous observation of the polar movement. In Japan, there is one in Mizusawa, Iwate Prefecture. Every five days, survey data from 80 observations is sent to the International Latitude Observation Bureau (B.I.H.) in France.

It has been reported in recent years that the polar movement has been deviating abnormally from the almost uniform Chandler Cycle. To illustrate this, I arranged the data obtained from B.I.H. into a graph (Figure 2). The accuracy of this data is within one hundredth of a second, based on the unit of angle, which is excellent indeed. From the graph you can see that movement is considerably abnormal and irregular.

Generally the polar movement has a natural flow. The flow, as described by the word "natural", usually moves in a smooth curve. The Chandler Cycle has been compared to the axial movement of a children's top, which moves in a circle even when tilted. But, as the graph shows, there are very unusual, sudden shock-wave changes of considerably acute angles. I must confess that I shuddered with horror at discovering this abnormality.

I then applied the dates of nuclear tests with a force of over one 150 Kilotons [TNT equivalent], as surveyed by the Foreign Ministry Information Department, to the Graph. Again, I found it obvious that the position of the pole slid radically at the time of a nuclear explosion.

However, there was movement which could not be explained by the nuclear test data alone. After information collected at the United States Geological Survey on world earthquakes with an intensity of over magnitude (M) 7 were plotted, the affect of the earthquakes on the polar movement was also clearly evident. On the graph, the effects of the nuclear tests and earthquakes on polar movement can clearly be seen.
(See Figure 2.)

As a result of the above study, a new study theme was born. Namely, the affect of nuclear tests on earthquakes. There have been 12 earthquakes with an intensity of over M7, from February to December, 1975. Six of them have occurred within 10 days after a nuclear test was conducted.

Generally, earthquakes are caused by an accumulation of stress in the Earth's crust. However, it can be said that there is a strong possibility of earthquakes being artificially triggered by atomic explosions.

Changes in the polar movement naturally cause the rotation cycle of the Earth to change. This, in turn, can affect time, which we have always considered to be innate and invariable. Even our 24 hour day might begin to be affected.

The 24 hour day is based on the period it takes for a certain star to pass over a fixed point on Earth and return to that fixed point. It has already been observed with high precision instruments that time is exact on some days,

continued on page 9.

Figure 2: Data from the international Latitude Observation Bureau is arranged chronologically in the graph. The pole of the Earth should rotate in a near circle (Chandler Cycle). On the contrary, the trace of the pole shows a zig zag movement.

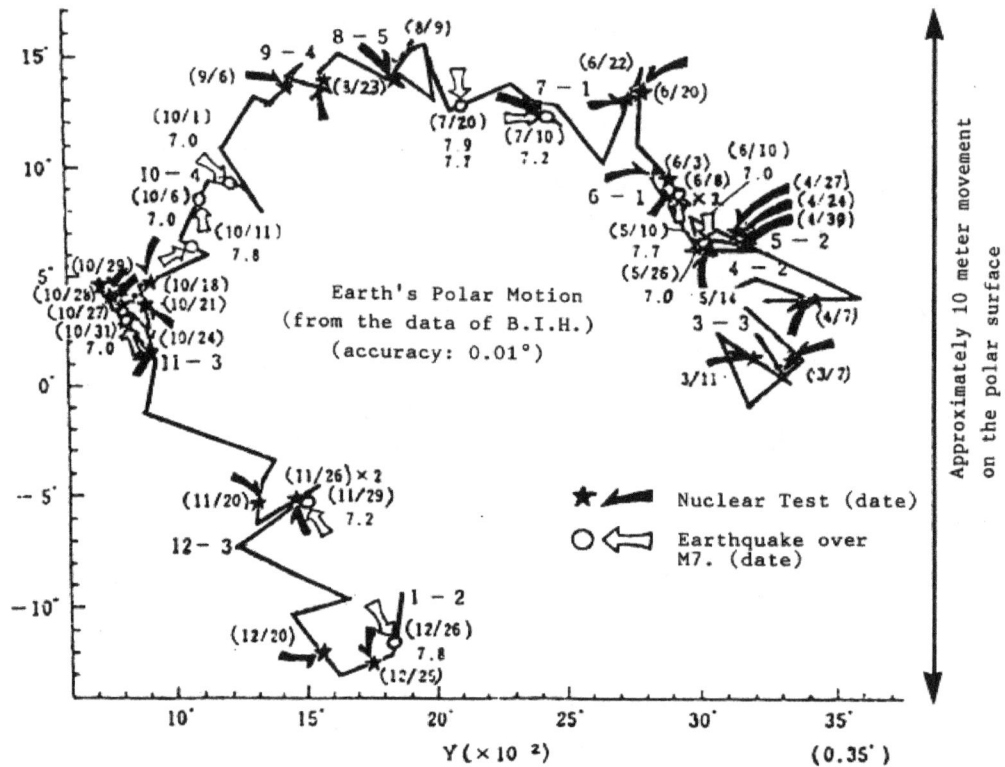

Figure 3. The rotation of the earth increases or decreases, and the affects of nuclear test and earthquakes are clearly evident in the variations of the normal wave curve.

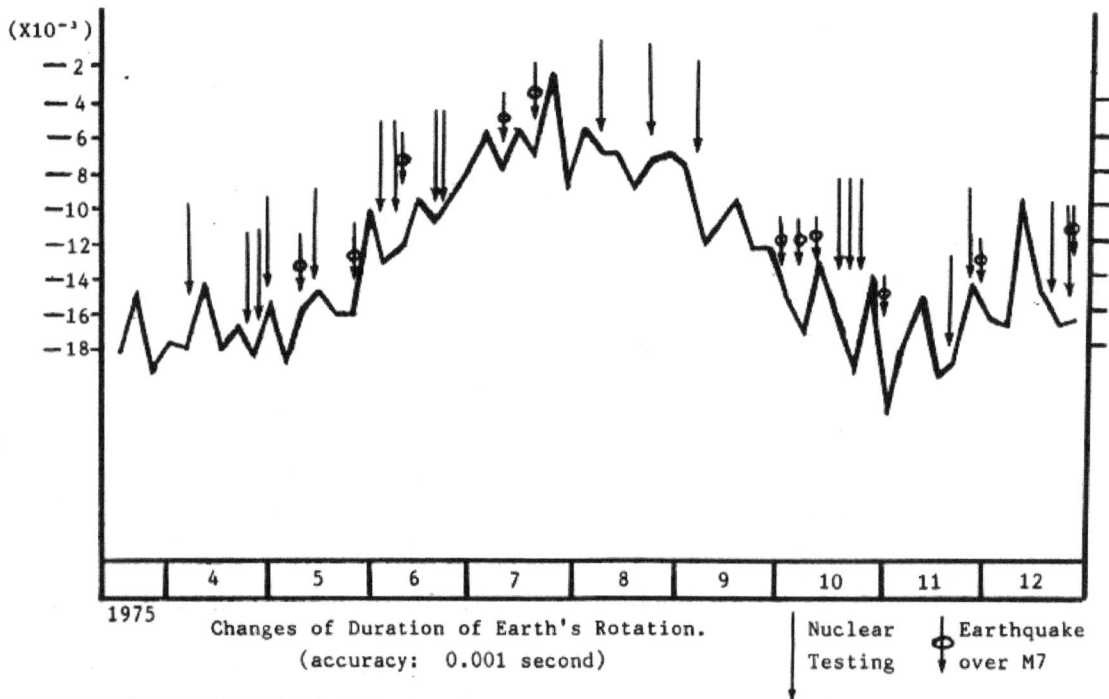

Changes of Duration of Earth's Rotation.
(accuracy: 0.001 second)

and slightly different on others. Even though the difference may be only a thousandth of a second, time has begun to deviate due to the nuclear testing (Figure 3).

Conclusion

Through this study, I began to reconsider the relationship between man-made disturbances in the natural environment, and the possibility of a limit to the endurance of nature against the challenges of man. The temperature of the atmosphere is changed by nuclear tests — a change that even the Sun cannot affect. One can easily guess how the nuclear explosions affect the meteorological conditions of the Earth.

Nuclear explosions move the very axis of the Earth. This is nothing short of flaunting the fear of God. Mankind cannot afford this kind of indifference.

In the face of these grave dangers, what lies ahead?

For example, changes in the force of charged particles and ultraviolet rays from the sun do not particularly disturb the normal atmospheric environment of the stratosphere. Yet [atmospheric] nuclear tests not only change the temperature of the stratosphere, they cause serious resultant disruption of the atmospheric circulation system.

From the information so clearly defined by our study of only a little over a year, I believe that nuclear testing must be stopped immediately, regardless of its peaceful or military purpose.

Recently, Typhoon #17 remained stationary in the open sea near Kyushu, Japan. This frightening phenomenon prompted some people to argue the possibility of using nuclear force to combat typhoons. This must never be permitted. It cannot be denied that our study has conclusively proven the large scale destruction of nature caused by nuclear explosions.

For Additional Information:

* Bolt, B.A.: *Nuclear Explosions and Earthquakes, the Parted Veil*, W.H. Freeman, San Francisco, 1976.
* Dahlman, O., et al: "Ground Motion and Atmospheric Pressure Waves from Nuclear Explosions in the PolynesianTest Area Recorded in Sweden, 1970", FOA 4 Report, C4461-26, 1971
* Davidson, C.I., et al: "Radioactive Cesium from the Chernobyl Accident in the Greenland Ice Sheet", *Science* 237:633-634, 7 August 1987.
* Dibble, T.C.: "H-Bombs Have Us Quaking", private publication, 1959 (reprinted 1987 by *Borderland Sciences,* PO Box 429, Garberville, CA 95440).
* Harris, D.L.: "Effects of Atomic Explosions on the Frequency of Tornadoes in the US", *Monthly Weather Review,* December 1954, p.360-369.
* Kato, Y. & Matsumae, S.: "Recent Abnormal Phenomena on Earth and Atomic Power Tests", Tokai University, *Pulse of the Planet,* 1(1):4-9, 1989.
* "Lightning Increase After Chernobyl", *Science News,* 312:238, 10 October 1987.
* Reich, W.: *The Oranur Experiment, First Report (1947-1951),* Wilhelm Reich Foundation, Maine, 1951 (partially reprinted in Reich, W.: *Selected Writings,* Farrar, Straus & Giroux, 1960).
* Smith, R.J.: "Scientists Implicated in Atom Test Deception", *Science,* 218:545-547, 5 November 1982.
* Smith, R.J.: "Atom Tests Leave Infamous Legacy", *Science,* 218:266-269, 15 October 1982.
* Trombley, A: "An Interview...", *Wildfire,* p. 17-33, January, 1988.
* Matsushita, S., et al: "On the Geomagnetic Effect of the Starfish High Altitude Nuclear Explosion", *J. Geophys. Res.* 69:917-945, 1964

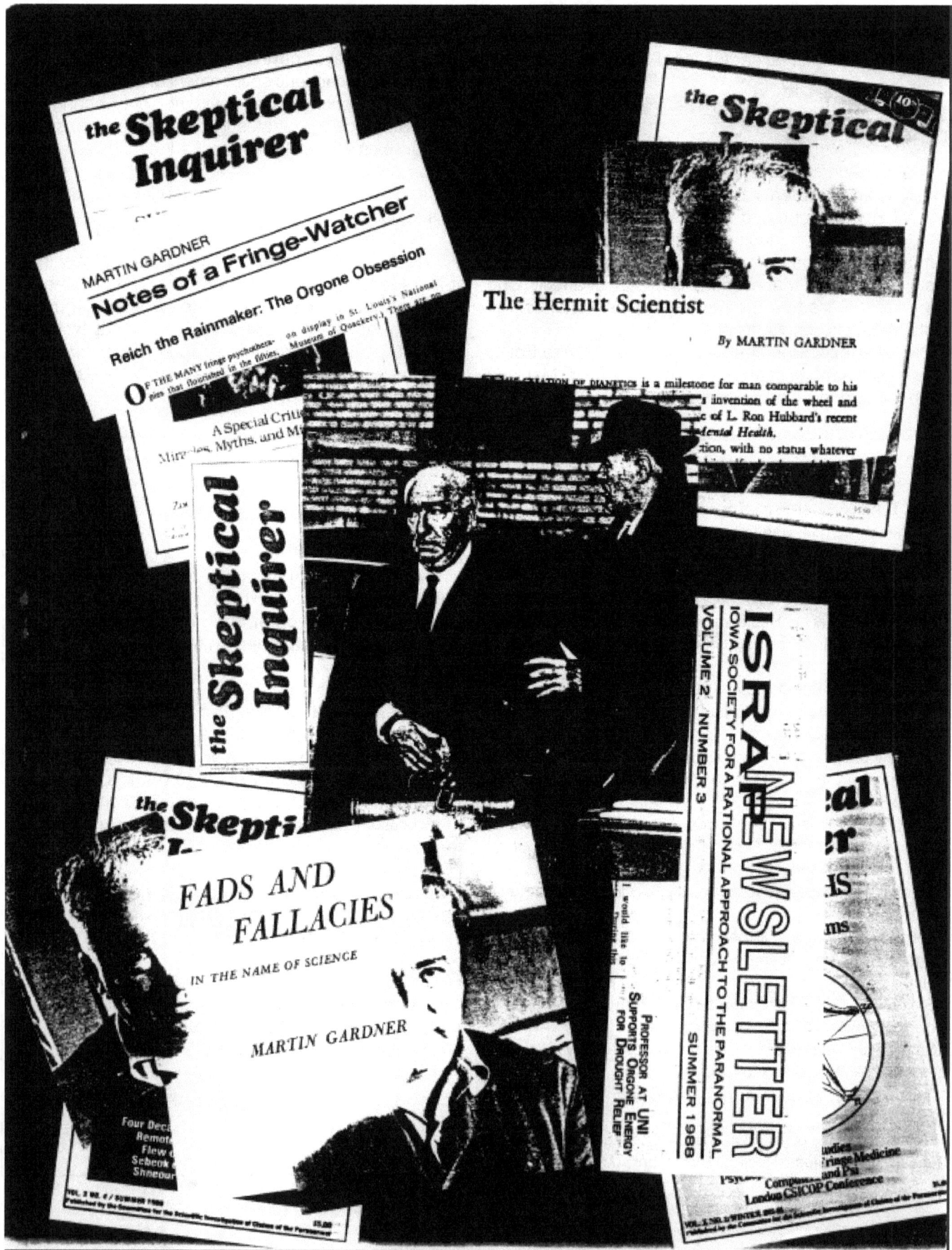

Response to Martin Gardner's Attack on Reich and Orgone Research in the *Skeptical Inquirer*

James DeMeo, Ph.D.*

The journal *Skeptical Inquirer* is the official publication of CSICOP, the "Committee for the Scientific Investigation of Claims of the Paranormal". The organization has a reputation for debunking many popular beliefs of either a metaphysical or folk-lore nature. Among their favorite targeted subjects are astrology, ESP, UFOs, psychokinesis, faith-healing, and psychic surgery. CSICOP has made the headlines in recent years for its attacks on advocates of "paranormal" phenomena, and for actual unmasking of a few deceptive "faith healers". But its membership has also expressed opposition to any unusual ideas that do not fit within a very narrow, mechanistic world view, such as solar-terrestrial correlations, acupuncture, and dietary treatments for degenerative disease. In recent months, the organization was itself publicly tarnished following their attack upon Jacques Benveniste, a French scientist whose experiments provided some evidence for the principle of homeopathic dilutions.(1) Given their apparent reluctance to rely upon fair and open discussion, or honestly-conducted research as a means of resolving scientific controversies, CSICOP has since been labeled the "Truth police", "science cops", and other names by various members of the scientific community.

Most recently an article attacking Wilhelm Reich and orgonomy, by CSICOP leader Martin Gardner, appeared in the *Skeptical Inquirer*.(2) Titled "Reich the Rainmaker: The Orgone Obsession", the article takes aim at Reich primarily for his discovery of the orgone energy. To Gardner, Reich was a man gone mad, a "paranoid egoist". In the article, Gardner also recounts a small bit of my own research with the cloudbuster, which he attempts to condemn via association with the distorted picture of Reich he has painted. The article reeks with contempt for Reich, and for the whole concept of energy in space, and contains so many falsehoods, distortions, and half-truths that rebuttal requires some lengthy documentation. Only someone unread about the facts of Reich's life and works will find Gardner's article convincing.

* Director of Research, Orgone Biophysical Research Lab, PO Box 1395, El Cerrito, CA 94530 USA

Gardner mentions a few of Reich's research findings, but in such a manner as to invite disbelief, without any attention to details, or mention of the specific experiments which led to his conclusions. The article makes cartoons out of serious experimental work, and Gardner calmly asserts that the orgone is "an energy no physicist outside orgonomy circles has detected." This is quite a bald statement, but is completely false. Many examples will be given below of researchers who made little or no mention of Reich, who often strongly disliked him and the whole notion of the orgone energy, but who nevertheless unexpectedly detected an unusual, orgone-like energy in living creatures, in the atmosphere, or in space. First, however, let us briefly review what evidence has been gathered by Reich and his coworkers on the orgone question. I must reject Gardner's attempt to place automatically anyone who obtains positive evidence for the orgone within a suspect (and non-existent) "orgonomy circle". This is a dishonest attempt to cast suspicion and a taint upon anyone who actually does obtain positive evidence favoring Reich's claims. It is a method of ostracism common to cliques of schoolchildren on the playground, but has no place in scientific investigations. Furthermore, there has never been, to the best of my knowledge, any researcher who has ever carefully reproduced Reich's experiments and obtained clearly negative findings. Even Einstein confirmed one of Reich's experimental findings, the temperature differential within the orgone accumulator,(3) but unfortunately without completing the necessary control tests which demonstrate its orgone-energetic origins. Indeed, there are dozens of qualified researchers who have duplicated Reich's experiments, obtained positive confirming evidence, and published their findings in various journals.

Several years ago I produced a detailed *Bibliography on Orgone Biophysics*,(4) which covered the period of research from 1934 to 1986. It contains over 400 separate citations by more than 100 different authors, most of whom possessed the M.D. or Ph.D. degree. Besides my own thesis and doctoral dissertation,(5) which were presented to and accepted by a group of respected scholars at the University of Kansas, I have listed in this

" Does Gardner, a master with math games, card tricks, and use of the English language, have any research training or credentials to support his self-proclaimed authority over this matter?"

Bibliography 17 other theses and dissertations which drew heavily from Reich's works, confirming various aspects of his bioenergetic formulations. There are 38 indexed citations in the *Bibliography* covering Reich's bion and biogenesis experiments, including Professor du Teil's 1938 confirming presentation on the bions to the French Academy of Sciences. The *Bibliography* also contains more than 80 indexed citations on the electro-scopical, thermical, and biological effects of the orgone energy accumulator. This includes some 22 studies on plant-growth responses, and 6 on cancer retardation or wound-healing in laboratory mice. Another 12 citations discuss or evaluate the Reich bioenergetic blood test. More than 50 citations focus on cloudbusting, with 20 or so papers discussing methods for direct visual observation of the atmospheric orgone. Of particular note is the most recent German dissertation on "The Psycho-Physiological Effects of the Reich Orgone Accumulator",(6) which was a double-blind, controlled study, confirming many details of Reich's original assertions on the parasympathetic stimulation of concentrated orgone energy on the body, and the weather-dependent pulsation of the orgone in the accumulator.

But Gardner says nothing about this research, as if it was wothless, the workers involved being somehow deluded into forgetting their research training, or worse. I ask, can he specifically cite anyone, even a single person, who has duplicated any one of Reich's experiments and obtained a fully negative result? Has he ever personally attempted to reproduce a single one of Reich's experiments, or even the more simple observational tests? Can he demonstrate even a cursory knowledge of this body of positive research evidence, which extends back some 50 years, or give a convincing, rational reason for his contrived and easy dismissal of it all? Does Gardner, a master with math games, card tricks, and use of the English language, have any research training or credentials to support his self-proclaimed authority over this matter? Does he not care a whit for the facts in his overwhelming drive and passion to skewer Reich and the orgone? The answer appears to be *NO* on all counts.

What Gardner fails to mention is also telling. For example, one would not know from his article that Reich had his books and research journals *banned and burned* by an American court of law, with five actual episodes of court-ordered book-burning taking place, most recently in the 1960s. He mentions the fact that Reich was, in 1932 and 1933, disowned by both the International Psychoanalytic Association (IPA) and the German communists, but failed to mention that he was likewise attacked, and put on

death lists, by both the Nazis and Stalinists, who also burned his books. After fleeing from Hitler's Germany, Reich was welcomed by the Norwegian analysts, who liked his writings, and disagreed with the politically-motivated actions of the IPA.(7) But Gardner is not concerned with details, as he considers Reich's work, and that of his coworkers, to be "religion". He compares orgonomy, which makes no claim to metaphysical truths or salvation, and has no gurus, churches, and the like, to *Scientology*, a self-proclaimed religion with churches, sunday services, and a messianic leader widely know for his science-fiction writing. One would not know, for example, that orgonomy is a research discipline developed from new natural scientific observations and experimental findings.

Gardner's first attack against Reich appeared in the *Antioch Review* of 1950,(8) though he was then more restrained in his linguistic distortions and vituperation. In 1952 he attacked Reich, with similar clever wit and fervor, in a chapter in *Fads and Fallacies in the Name of Science*.(9) His articles helped fuel the Food and Drug Administration's (FDA) pseudo-investigation, which has since been demonstrated, through at least three different Freedom-Of-Information-Act searches of FDA files,(10) to have been conducted in a most shabby, antiscientific "get Reich" manner. Today, we know that there is no

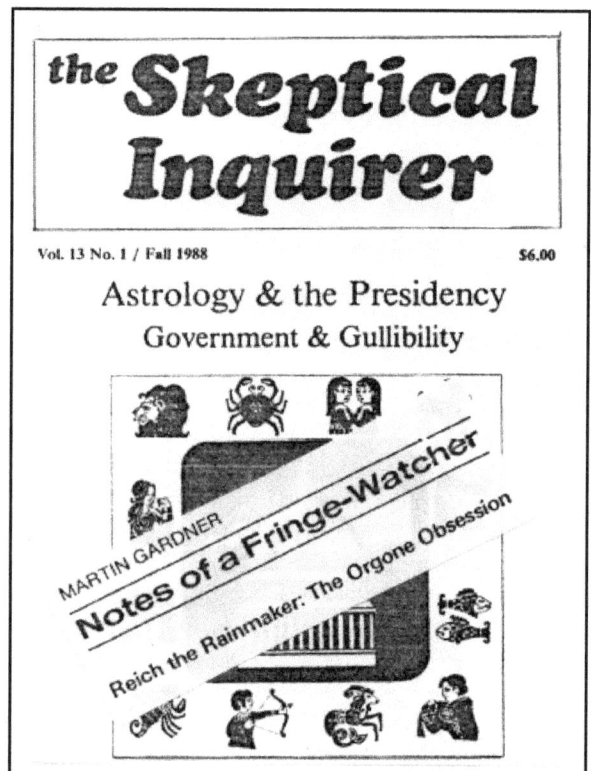

the *Skeptical Inquirer*

Vol. 13 No. 1 / Fall 1988 $6.00

Astrology & the Presidency
Government & Gullibility

MARTIN GARDNER
Notes of a Fringe-Watcher
Reich the Rainmaker: The Orgone Obsession

> **" The thought that these younger orgonomists might have been persuaded to accept Reich's findings by *weight of evidence* does not pass into Gardner's pen."**

credible evidence contained in FDA files by which they could have justified their actions.(See the article on Page 18 in this issue of the *Pulse*.)

Reich, of course, was outraged that various hack journalists had slandered him, and put false words into his mouth about the effects of the accumulator.(11) Gardner incautiously repeats some of these falsehoods in his recent article, such as "The concentrated orgone is said to relieve symptoms of almost every illness from cancer to impotence."(2:26) He was more cautious in his earlier articles on Reich. In the *Antioch* article he also asserted that no competent scientist would bother to refute Reich's findings, condemning them with a wave of the hand. Gardner was obviously wrong in that Reich has not been ignored since the 1950s, by either scientists or laypeople. But Gardner continues to deny and ignore the *experimental, empirical nature of Reich's findings,* which have guaranteed a continuing, growing interest in them for over three decades. Having failed in his 30-year mission to distort the facts, Gardner's latest attack reveals a harsher, more frustrated tone. Interestingly, in his early articles the younger Gardner made at least passing mention of *Lysenko,* a Stalinist bureaucrat who put many scientists to death for their research findings; but not so the elder Gardner, who has lost sight of the lessons of history, and seems glad that Reich died in jail, his books condemned to flames. I believe this is because Gardner, and other politically-powerful media-darlings of the CSICOP gang, have been quietly and consistently asserting a deadly new form of Lysenkoism in the USA, for at least 30 years. Is there anyone who would deny the fact that academic freedom is almost non-existent in the USA if one wishes to seriously study certain questions, such as the orgone energy, or, for that matter, anything which challenges the assertions of "empty space", "every cell from a cell", or non-genetic mechanisms for heredity? My files grow increasingly full of recent examples of American researchers and medical pioneers who have been trounced into silent submission, into jail, or prematurely into their graves, for doing nothing more than exploring such questions!

CSICOP claims, on the back cover of its publication, that it investigates "fringe-science claims from a responsible scientific point of view", and also does "not reject claims on a-priori grounds antecedent to inquiry, but rather examines them objectively and carefully". However, from the above, we have seen that the Gardner article, at least, has violated these high-sounding goals in an extreme way. The detailed scientific research of Reich and his coworkers is flippantly ignored, as if it does not

really constitute "research", their experiments somehow failing to be real "experiments". But he, Gardner, writes as if he had examined all the facts and evidence, when the truth is that he has done little or no examining at all, other than to select quotes cleverly here and there from a few books. CSICOP and Gardner have set a pattern for themselves. They proclaim expertise over matters where they have none, and condemn it where it exists. It is not so much different from the 1950s, when the FDA substituted rumor and gossip for "evidence", granted "expertise" only to those scientists who had demonstrated the proper quanta of ignorance, contempt, and prejudice, and concocted "experiments" which bore no resemblance to those previously published.

Dr. Reich's findings have not died with him because his experiments, when carefully conducted under the original conditions, produce the same results now as when he first developed them. They yield clear evidence for a pulsatory, weather-active and biologically-active energy continuum. It can, and has been, measured and photographed, and found to exist in high vacuum as well.(4) Reich called this energy continuum the orgone, but other scientists, working completely independent of Reich, and usually without knowledge of his works, have likewise measured or strongly inferred the existence of such an energy.

For example, there is Dayton Miller's work on the dynamic aether drift,(12) Halton Arp's work on energy/matter bridges between galaxies in deep space,(13) Giorgio Piccardi's work on solar influences upon the physical chemistry of water,(14) Frank Brown's work on cosmic modulation of biological clocks,(15) Harold Burr's work on the electrodynamic characteristics of creatures and the natural environment,(16) Hannes Alfven's work on streaming plasmas in the depths of space,(17) Thelma Moss' work on energy-field photography,(18) Bjorn Nordenstrom's work on x-ray phantom-images and circulation of bioenergy,(19) Robert Becker's work on mammalian bioelectrical limb regeneration,(20) Rupert Sheldrake's work on morphogenetic fields,(21) Louis Kervran's work on bioenergy-driven biological transmutations,(22) Berkson, Emergy, Anderson and Spangler's works on non-constant, continuum effects in nuclear decay processes,(23) and Paul Dirac's observations on the "neutrino sea".(24) And yes, we must not forget the work of CSICOP target Jacques Benveniste,(1) who demonstrated a non-molecular, likely *energetic* phenomena long known to homeopathic physicians. Each of these workers discovered or argued for a force conceptually similar to orgone: mass-free, yet capable of affecting or being bound to

> **" Mystical physics of today says we can't possibly touch or see these forces directly, given that they expended their influence billions of years ago."**

matter, participating in physical chemistry, metabolism, and heredity in some way, possessing measurable biological, meteorological, and cosmic components, reflectable by metal shielding, yet also amplifiable (and not extinguishable) through use of solid metal enclosures. Only in the case of Moss do I recall orgone being mentioned as a possible mechanism, but the properties and behavior of the phenomena independently identified by these researchers were orgone-like in many ways. So much for the assertion that no one "outside orgonomy circles" has detected these phenomena.

Other aspects of Gardner's attack on Reich focus upon his personal life, and his observations of UFOs. Here, highly selective quotes and exaggerations paint an awful portrait of Reich; one would never guess that he was admired by his coworkers for being an emotionally honest, patient, and gentle man. But is this really an issue? Could we condemn the telephone or the light bulb if it were proven that Bell or Edison somehow behaved in an "unlikable" way? Reich's personal life has no bearing at all on the question of whether or not the accumulator or cloudbuster really function as described; similarly regarding UFOs, which Reich, and a host of other reliable witnesses have seen from time to time. Unless we wish to focus specifically upon the question of UFOs, or upon Reich's Arizona experiments, his *speculations* about the nature of the UFO are of only passing interest.

Gardner is also very selective when discussing the childhood recollections of Reich's son, Peter.(25) He cites the passage where Peter Reich helps his father work the cloudbuster in the Arizona desert, where UFOs were observed, but says nothing of the child's recollections of Government Agents invading his father's Maine laboratory, putting accumulators to the axe, and carrying away crates of books for burning in incinerators. And likewise regarding Reich's writings in *Contact With Space*.(26) The UFO observations are mentioned again and again, but nothing is said of Reich's successful experiment for bringing moisture to the deserts. Gardner brands Reich paranoid for his speculative, and forlorn writings from this time when he was working almost entirely alone in the middle of the desert, and under malicious attack by the popular media, by academics, and by the Government. The diagnosis of "paranoia" is only correct in circumstances where there is no real threat to the individual in question, only a falsely perceived one. In Reich's case the threats were real.

Gardner's discussion of my work was in many cases either only partly true or incorrect. This may be due to the fact that he relied on dubious sources for informa-

tion on my research. Rather than write me for details on my published experiments,(5, 27) Gardner relied on a *National Enquirer* article I know nothing about. He also relied on second-hand, word-of-mouth recollections of a lecture I gave to the Association for Arid Lands Studies,(28) or a clandestinely recorded version of that lecture. I learned about the *Skeptical Inquirer* article not from its editors, from CSICOP, or Gardner, but via an anonymous phone call. Such intrigue! Whatever, the figures Gardner cited, of my engaging in 13 successful cloudbusting operations out of 15, are several years out of date. As of October 1988, I have participated in or directed over 30 different cloudbusting operations, more than half of which took place during mild to severe drought conditions, or desert conditions. Approximately 80% of these operations were successful in that significant rains, and other major and distinct atmospheric changes, developed within 48 hours after onset of operations. This success rate is preserved for the drought-desert operations, as assessed independently. These latter experiments include work during the 1986 Southeastern drought, which did dramatically end shortly after our cloudbusting operations began, and a most recent cloudbusting operation (mid-September 1988) in the drought zone of the Pacific Northwest. Another successful cloudbusting operation recently took place in the harsh deserts north of Yuma, Arizona, confirming the desert-greening possibilities raised by Reich in *Contact with Space* over 30 years ago.* For the record, all my cloudbusting operations since 1980 have been preceded by a documentary telegram to NOAA (National Oceanographic and Atmospheric Administration). They are documented and evaluated via ground photos, satellite imagery, and National Weather Service data.

Gardner implies that I make a lot of money from the orgone research, when the truth is that it costs me a lot of money, and returns nothing financially. It costs thousands of dollars to launch a cloudbusting operation of any magnitude, and because of the hatred towards Reich's works which currently exists, funding is not available through ordinary sources. The *orgonotester*, which I import but do not manufacture, is made by the *Marah SA* company, which also makes top-notch air ion measuring equipment. Apparently Dr. Walter Stark, the Swiss ion expert who developed these instruments, is also interested in orgone energy.

I started giving workshops on "The Bioenergetic,

---

* See the May 1989 issue of the *Journal of Orgonomy* for details on these recent operations.

Orgonomic Basis of Life and Weather" after I saw the need for factual education on these subjects. For the record, the workshops are attended by many enthusiastic young students, often the brightest and best, who usually already know a lot about Reich, and feel deeply offended that so many of their professors put him down without the slightest notion of what they are talking about.

So we come to a point of consideration, namely why it is that Gardner attacks Reich in such a blind way? He is decidedly upset that the orgone question did not lay down and die with Reich in 1957:

> "One might have thought that today's orgonomists...would confine themselves to Reich's youthful contributions to psychoanalysis, which are reasonably sane and still greatly admired by many psychiatrists, but no — most of them buy it all."(2:28)

The thought that these younger orgonomists might have been persuaded to accept Reich's findings by *weight of evidence* does not pass into Gardner's pen. However, a clue to his own motivations for attacking and distorting the record is found in Gardner's own writings. In *The Whys of a Philosophical Scrivener*,(29) Gardner finally makes known his own world view. And what do we see? One reviewer puts it so:

> "...not what we might expect from an apostle of the rational. Gardner announces that he believes in the existence of God — not the pantheistic God of Spinoza or Einstein, but an omniscient creator who would be recognizable to anyone immersed in the Judaeo-Christian tradition. Gardner is deeply convinced of the possibility of a soul and an afterlife, if not of a conventional heaven and hell. He writes movingly about the benefits of prayer, not merely for its possible psychological value, but also because God might actually heed it."(30)

Now, Mr. Gardner is fully entitled to believe whatever he wishes, but we must note that Reich's *functional,* bioenergetic works stand in clear opposition to both a dead, machine-like universe, and a dualistic, "spirit-versus-flesh" anthropomorphic deity. Indeed, Reich argued persuasively that the mechanistic-mystical world view was the result of a perceptive splitting-off of organic sense functions, caused by the chronic damming-up of

emotional-sexual energy within the body of the observer.(31) For these reasons, he argued, *animistic* peoples, who lived a more vibrant and uninhibited emotional and sexual life, and who consequently remained relatively free of neuroses,(32) could feel, with their sense organs, the tangible energetic forces which shaped and created the universe. To them, the spirit-forces were dynamic, alive, in the "here and now", and not divorced off into some intangible "heaven" or "hell". Reich also pointed out the essentially mystical nature of many concepts of modern physics, wherein, like deistic religion, the basic forces which shape and structure the universe, are also *not* tangible, *not* directly measurable, and *not* observable through the senses. Mystical physics of today says we can't possibly touch or see these forces directly, given that they expended their influence billions of years ago, or are woven into the fabric of an unobservable "space-time continuum". According to this view, the central creative event which put the whole universe into motion occurred in a primordial "big-bang", which only by "accident", we are told, conforms to the biblical Genesis. This point of view might be convincing were it not for the fact that plenty of contrary empirical evidence exists. In addition to the evidence cited above, we may ask: what do Reich's functional discoveries do to such a world view?

Reich's orgone is a spontaneously pulsatile, excitable, and negatively-entropic energy. It is an active, creative principle which is tangible, real, measurable, and in the "here and now". Through experiment, it was found that concentrated, excited orgone in high vacuum absorbs and diminishes electromagnetic excitations transmitted through it. As such, it provides a mechanism for the red-shifting of galactic light, through a means other than doppler effects.(13, 31) These findings completely undermine the theoretical basis of the "expanding universe", the "big-bang", "relativity", and popular notions such as "black holes", etc. Indeed, any astrophysical theory which requires a constant light speed and "empty space" is undone by Reich's findings. And if Reich is correct about the streaming, pulsatile, superimposing nature of the orgone continuum in space,(33) it would also fulfill the requirements of *prime mover,* putting the anthropomorphic deity into the unemployment lines, and preserving Genesis only as historical literature, and not important natural philosophy. Who will deny the growing speculative tendency in certain quarters of astrophysics for linkage between the big-bang and the book of Genesis? This connection has not even been lost on the Pope!

But there is more. Reich also argues that the spontaneous aspects of life, namely those governing emotion and sexuality, are not only natural and biologically necessary, but also measurable and tangible.(31, 34) Sex is not a sin to Reich, and Original Sin is psychopathological myth. The sexual impulse is not intrinsically devilish but an aspect of bioenergetic superimposition and charge, striving for natural release, even among adolescents and the unmarried.

All this harkens back to a similar, historically important difference between the world views of Galileo, history's greatest empiricist, and Newton, a man who was preoccupied with theology. Galileo looked to the energetic aether as probable prime mover, at work in the here and now. He was antagonistic towards "revealed truth", and demanded that his critics reproduce his experiments before making judgments, to "look into the telescope". But not so Newton, who argued for dominance of the Church over matters of experimental science.(35) He proclaimed the aether to be static and immobile, without a shred of evidence in order to eliminate its participation in the ordering and movements of the Heavens. That role, he believed, belonged only to the Christian anthropomorphic God. Newton's theological restraints on scientific inquiry have remained to this day, and are even championed by a scientistic community bent on a near total denial of the bioenergetic in the natural world. In the late 1800s, Michelson and Morley searched for but did not detect Newton's *static* aether. But their student, Dayton Miller, did detect and fully document a *moving, dynamic, metal reflectable* form of it.(12) And so did Reich, who discovered this same dynamic energy as the sensible and measurable sexual-biological-cosmic orgone energy.(4) Reich's works not only undermine many popular "facts" regarding human behavior and the origins and functioning of life, but also all the various mechanistic and mystical theories of science which demand the absence of a dynamic energy in the natural world. Gardner and the CSICOP gang oppose Reich not because he failed to provide good empirical evidence for such an energy, but for just the opposite reason, *because he did.*

"When a true genius appears, you can know him by this sign: that all the dunces are in a confederacy against him." Jonathan Swift

REFERENCES:

1) E. Davenas et al, *Nature*, 333:832, 1988; J. Maddox, et al, *Nature*, 334:287, 1988; J. Benveniste, ibid, p.2; J. Benveniste, *Science,* 241:1028, 1988.

2) M. Gardner, "Reich the Rainmaker: the Orgone Obsession", *Skeptical Inquirer,* 13(1):26-30, Fall 1988.

3) W. Reich, *The Einstein Affair*, Orgone Institute Press (xerox avail. from Wilhelm Reich Museum, PO Box 687,Rangeley, Maine 04970), 1953.

4) J. DeMeo, *Bibliography on Orgone Biophysics*, Natural Energy Works (PO Box 1395, El Cerrito, CA 94530), 1986.

5) J. DeMeo, "Preliminary Analysis of Changes in Kansas Weather Coincidental to Experimental Operations with a Reich Cloudbuster", U. of Kansas thesis, Geography-Meteorology Department (xerox avail. from Natural Energy Works, PO Box 1395, El Cerrito, CA 94530), 1979; J.DeMeo, "On the Origins and Diffusion of Patrism: the Saharasian Connection", U. of Kansas dissertation, Geography Department (xerox avail. from University Microfilms), 1986.

6) S. Muschenich & R. Gebauer, "Die (Psycho-) Physiologischen Wirkungen des Reich'schen Orgonakkumulators auf den Menschlichen Organismus", U. of Marburg (FR ofGermany) dissertation, Psychology Dept. 1986. (Published as *Der Reichsche Orgonakkumulator,* Nexus Press (avail. through Natural Energy Works, PO Box 1395, El Cerrito, CA 94530) 1987.

7) M. Sharaf, *Fury on Earth, a Biography of Wilhelm Reich,* St. Martin's-Marek, NY, 1983.

8) M. Gardner, "The Hermit Scientist", *Antioch Review*, Winter 1950-1951, pp.447-457.

9) M. Gardner, chapter on "Orgonomy" in *In the Name of Science* (later titled *Fads and Fallacies in the Name of Science),* Dover, NY, 1952.

10) R. Blasband, "An Analysis of the United States Food and Drug Administration's Scientific Evidence Against Wilhelm Reich, Part 1: the Biomedical Evidence", *J. Orgonomy*, 6(2):207-222, 1972; C. Rosenblum, ..Part 2: the Physical Concepts", *J. Orgonomy,* 6(2):222-231, 1972; C. Rosenblum, ...Part 3: Physical Evidence", *J. Orgonomy*, 7(1):92-98, 1972; J. Greenfield, *Wilhelm Reich Versus the USA,* W.W. Norton, NY, 1974; J. DeMeo, "Postscript on the Food and Drug Administration's Scientific Evidence Against Wilhelm Reich", *Pulse of the Planet*, 1(1): 18-23, 1989.

11) J. Greenfield, ibid.; T. Wolfe, *The Emotional Plague Versus Orgone Biophysics, the 1947 Campaign,* Orgone Institute Press, NY, 1947; W. Reich, *Listen, Little Man,* Farrar, Straus & Giroux, NY, 1974.

12) D. Miller, "The Ether-Drift Experiment and the Determination of the Absolute Motion of the Earth", *Reviews of Modern Physics*, 5:203-242, 1933.

13) H. Arp, et al, *The Redshift Controversy*, W.A. Benjamin, Reading, MA 1973; H. Arp, *Quasars, Redshifts, and Controversies,* Interstellar Media, Berkeley, CA, 1987; cf. C. Rosenblum, "The Red Shift", *J. Orgonomy*, 4:183-191, 1970.

14) G. Piccardi, *Chemical Basis of Medical Climatology*, C. Thomas, Springfield, IL, 1962; cf. J. Bortels, "Die Hypothetische Wetterstrahlung als vermutliches Agens Kosmo-Meteoro-Biologischer Reaktionen", *Wissenschaftliche Seitschrift der Humboldt-Universitat zu Berlin,* VI:115-124, 1956.

15) F. Brown, "Evidence for External Timing in Biological Clocks", in *An Introduction to Biological Rhythms*, J. Palmer, ed., Academic Press, NY, 1975.

16) H. Burr, *Blueprint for Immortality*, Neville Spearman, London, 1971; cf. L. Ravitz, "History, Measurement, and Applicabil-

ity of Periodic Changes in the Electromagnetic Field in Health and Disease", *Annals, NY Academy of Sciences*, 98:1144-1201, 1962.

17) H. Alfven, *Cosmic Plasmas*, Kluwer, Boston, 1981; cf. , "The Big Bang Never Happened", *Discover,* June, 1988, pp.70-80.

18) T. Moss, *The Body Electric: A Personal Journey Into the Mysteries of Parapsychological Research, Bioenergy, and Kirlian Photography*, J. P. Tarcher, Los Angeles, 1979.

19) B. Nordenstrom, *Biologically Closed Electric Circuits: Clinical, Experimental and Theoretical Evidence for an Additional Circulatory System,* Nordic Medical Publications, Stockholm, Sweden, 1983.

20) R. Becker & G. Selden, *The Body Electric: Electromagnetism and the Foundation of Life*, Wm. Morrow, NY 1985.

21) R. Sheldrake, *A New Science of Life, The Hypothesis of Causative Formation,* J. P. Tarcher, Los Angeles, 1981.

22) L. Kervran, *Biological Transmutations,* Beekman, Woodstock, NY, 1980.

23) J. Berkson, "Examination of Randomness of Alpha Particle Emissions", *Research Papers in Statistics*, F.N.David, ed., Wiley, NY, 1966; G. Emery, "Perturbation of Nuclear Decay Rates", in *Annual Review of Nuclear Science,* Annual Reviews, Palo Alto, CA 1972; J. Anderson and G. Spangler, "Serial Statistics: Is Radioactive Decay Random?", *J. Physical Chemistry,* 77:3114-3121, 1973

24) P. Dirac, "Is There An Ether?", *Nature,* 162:906, 1951; also see L. deBroglie, *Non-Linear Quantum Mechanics,* Elsevier, NY, 1960; H. Dudley, *New Principles in Quantum Mechanics,* Exposition University Press, NY, 1959, H. Dudley, *Morality of Nuclear Planning,* Kronos Press, Glassboro, NJ, 1976. I. Asimov, *The Neutrino,* Avon Books, NY, 1966.

25) P. Reich, *A Book of Dreams,* Harper & Row, NY, 1973.

26) W. Reich, *Contact With Space,* Core Pilot Press, NY, 1957.

27) J. DeMeo, "Field Experiments with the Reich Cloubuster: 1977-1983", *J. Orgonomy,* 19(1):57-79, 1985; J. DeMeo & R. Morris, "CORE Progress Report #13, Fighting the Extreme Drought of Spring 1985: Southeast", *J. Orgonomy,* 19(2):265-266, 1985; J. DeMeo & R. Morris, "CORE Progress Report #14: Possible Slowing and Warming of an Arctic Air Mass Through Cloudbusting", *J. Orgonomy,* 20(1):120-125, 1986; J. DeMeo & R. Morris, "CORE Progress Report #15: Breaking the 1986 Drought in the Eastern U.S., Phase 3: A Cloudbusting Expedition into the Southeastern Drought Zone", *J. Orgonomy,* 21(1):27-41, 1987; J. DeMeo & R. Morris, "Preliminary Report on a Cloudbusting Experiment in the Southeastern Drought Region, August 1986", *Southeastern Drought Symposium Proceedings,* March 4-5, 1987, Columbia, SC., South Carolina State Climatology Office Publication G-30, pp.80-87, 1987.

28) J. DeMeo, "Nine Years of Field Experiments with a Reich Cloudbuster: Positive Evidence for a New Technique to Lessen Atmospheric Stagnation and Bring Rains in Droughty or Arid Atmospheres", *Abstracts of Papers,* Program of the 1987 Meeting of the Association for Arid Lands Studies, El Paso, Texas, p.6, 1987.

29) M. Gardner, *The Whys of a Philosophical Scrivener,* Quill, NY, 1983.

30) F. Golden, Book Review, *Discover,* October 1983, pp.88-91.

31) W. Reich, *Ether, God and Devil,* Farrar, Straus & Giroux, NY, 1973.

32) B. Malinowski, *Sexual Life of Savages,* Routledge & Keegan Paul, London, 1932; W. Reich, *The Function of the Orgasm,* Noonday, NY, 1971; W. Reich, *The Sexual Revolution,* Octagon Books, NY, 1971; V. Elwin, *The Muria and their Ghotul,* Oxford U. Press, Calcutta, 1947; J. Prescott, "Body Pleasure and the Origins of Violence", *The Futurist,* April 1975, pp.64-74; J. DeMeo, "On the Origins and Diffusion of Patrism: The Saharasian Connection", ibid.

33) W. Reich, *Cosmic Superimposition,* Wilhelm Reich Foundation, Rangeley, Maine, 1951.

34) W. Reich, *The Bioelectrical Investigation of Sexuality and Anxiety,* Farrar, Straus & Giroux, NY, 1982.

35) L. C. Stecchini, "The Inconstant Heavens" in *The Velikovsky Affair, The Warfare Of Science and Scientism,* A. deGrazia, Editor, University Books, NY, 1966; D. Kubrin, "How Sir Isaac Newton Helped Restore Law'n Order to the West", unpublished monograph, 1972.

NOTE: For an update on Dr. DeMeo's research since this early 1989 article, please review these websites:

* James DeMeo's Research Publications:
      https://orgonelab.academia.edu/JamesDeMeo

* Orgone Biophysical Research Lab:
      http://www.orgonelab.org

* Saharasia web page:
      http://www.saharasia.org

* Natural Energy Works, books:
      http://www.naturalenergyworks.net

Postscript on the Food and Drug Administration's Evidence Against Wilhelm Reich

James DeMeo Ph.D.*

Introduction

Since the 1957 death of Dr. Wilhelm Reich in prison, and the banning and burning of his books by the U.S. Food and Drug Administration (FDA), an unjustified cloud of suspicion continues, in the popular mind, to follow his name and work. Of those contemporary research scientists who have heard of Reich, or of the orgone energy he discovered, most display little or no interest, irrespective of the breakthrough nature of his findings. And while a growing number will express a positive interest in his works, all too many continue to be arrogant and dogmatic in their opposition. Only a few of Reich's critics have actually read or reviewed his published research findings, and none, to my knowledge, have bothered to honestly reproduce his experiments, following the original experimental designs and protocols. When asked to justify their negative opinions about Reich, these critics usually point to the investigation undertaken by the FDA in the 1950s, and Reich's subsequent troubles with the courts, as evidence of his incorrectness. This paper is focused upon that contention, namely what the FDA did or did not do with respect to evaluating Reich's research.

Since the late 1950s, political reform in the United States has progressed to the point where the average citizen has gained the right to access and review government documents and letters. The Freedom-Of-Information-Act (FOIA) has been used by different scholars to access the FDA's internal documents on the Reich case. To date, there have been at least three separate studies made of FDA files using FOIA file searches. These include the early 1970s study of Blasband and Baker,(1,2,3), and that of Greenfield.(4) This author likewise engaged in a FOIA search of FDA files, in the early 1980s.

* Director of Research, Orgone Biophysical Research Lab, El Cerrito, CA 94530 USA

The Blasband-Baker Study

Dr. Richard Blasband and Dr. Courtney Baker published "An Analysis of the FDA's Scientific Evidence Against Wilhelm Reich" (1,2,3). In this work, the authors discussed the results of their FOIA search of the original files and experimental records from the FDA investigation of Wilhelm Reich. The points listed below summarize the conclusions detailed in the original Blasband-Baker "Analysis" paper. In recounting these points, I stress that each was confirmed in my own independent FOIA search and review of FDA files and documents.

1) The experiments which Wilhelm Reich had developed to document and demonstrate a new phenomenon, the orgone energy, were not properly duplicated or controlled by the FDA; many of the experimental tests worked out by Reich were never duplicated at all. Some of the experimental findings developed by Reich were dismissed by FDA scientists following a textual, "armchair analysis". For instance, the electroscopical, thermal, and Geiger-Muller effects of the orgone energy accumulator, which had been detailed by Reich through a series of carefully controlled laboratory experiments, were simply dismissed out-of-hand by one FDA physicist after he reviewed a few of Reich's books.

2) Of those experiments which the FDA scientists did duplicate, Reich's original laboratory design and control procedures were generally ignored by lab workers who, apparently, had not done enough reading on the matter to know just what they were testing for. Only a few of Reich's orgone physics experiments were actually attempted, by Kurt Lion, a physicist working at the Massachusetts Institute of Technology (MIT), who focused on the thermal and Geiger-Muller (GM) effects of the orgone accumulator. In the case of the GM effects, positive evidence was directly observed by Lion, but misinterpreted. Lion's work also exemplified the problem of failure to adequately control the experiments, and duplicate them under the original laboratory conditions; some of these points regarding Lion's work will be discussed in

more detail below.

3) Regarding the medical testing of the orgone accumulator, the clinicians collecting observations and raw data often noted, but otherwise ignored, positive evidence for the medical effects of the orgone accumulator. The FDA administrators who oversaw these experiments also ignored this positive evidence, and emphatically maintained that no positive evidence whatsoever was observed. Raw data from the FDA clinical trials also reflected some effects of the orgone accumulator upon blood pressure and heart rate, as originally noted by Reich. However, these observations were likewise dismissed, as were some observed pain relief and accelerated healing effects of the orgone radiation upon burns; some of these effects were noted by the patients themselves and were entered into the clinical record, but were not duplicated in the FDA administrator's conclusion statements summarizing the clinical trials.

4) Regarding tests of the orgone accumulator on cancer mice, the clinicians involved deviated from Reich's experiments and used transplanted tumors in a very poorly controlled laboratory environment, where orgone-irritating x-ray machinery was nearby. Reich's experiments, by comparison, used mice genetically bred to allow spontaneous tumor development, and his tests were run in areas where high frequency devices, x-ray machines, TV sets, fluorescent lights, and radiation sources of all kinds were strictly excluded.

5) The FDA's cancer tests on humans with the orgone accumulator were restricted to terminally ill patients who had previously run a gamut of unsuccessful surgery, radiation, and chemotherapy. Some of these patients apparently were so feeble they had to be helped into the accumulator. Still, the FDA data might be interpreted as demonstrating an effect of the orgone accumulator upon a few of those patients. Blasband appropriately characterized these FDA efforts as "testing for miracles".

6) In general, the FDA experiments were structured so differently from Reich's and performed in environments so uncontrolled that, as Blasband and Baker concluded, they cannot be considered as valid attempts to duplicate his work.

The Greenfield Study

Additional evidence of FDA bias has been documented by Professor Jerome Greenfield, in his book "Wilhelm Reich Versus the USA".(4) Greenfield's research, which involved both interviews with FDA staff and FOIA file searches, revealed that the FDA's investigation initially had more to do with Reich's sexual theories than the orgone accumulator. Reich had been a thorn in the side of the traditional medical and psychiatric community, claiming as he did that neuroses and destructive aggres-

sion were created and supported by the prevailing anti-child, sex-negative, patriarchal social order. Modern medicine and psychiatry say little about these matters, but have maintained a constant support of the existing social order, from which their wealth and political power comes. Reich was one of the sharpest critics of this poisonous medical/psychiatric attitude, and the FDA was spurred on to "get Reich" on whatever charges it could. Much of the history of the alienation of Reich from the traditional medical community, and the use of gossip and slander against him in both medical journals and the popular media, has been detailed and discussed by Greenfield,(4) Wolfe,(5) and more recently in a biography of Reich, by Sharaf.(6) Reich's enemies had fed the FDA bureaucrats various slanders about a "sex cult" centered around his laboratory in rural Maine, and FDA investigators went snooping about, asking local people and various scientists very misleading questions. Reich's cancer research, which included tests on mice and humans with the orgone accumulator, became the focus of the FDA investigation only later on, after no evidence for the sexual slander materialized. During this period of harassment, a number of Reich's coworkers, and people who had a very peripheral association with his Maine laboratory, were fired from jobs in hospitals and public schools. Any connection with Reich was considered a reason for dismissal from a teaching, research, or medical post. (This plaguey situation continues to this day; I presently have in my files many documented cases of students, research scientists, and professors being harassed, kicked out of their programs, or fired for simply attempting to openly replicate Reich's experiments. Medical students will, to this very day, hide any interest they may have in Reich's works, and publish their findings under pseudonym.)

It should be noted that when the FDA first began to investigate Reich, he cooperated with them, and sent a letter to all his coworkers asking them to assist the FDA in whatever manner they could. However, when he learned that the FDA was looking for a "sex racket", he saw immediately that the "investigation" was a sham, became justifiably infuriated, and ceased cooperating. He demanded that the FDA explain themselves, and he asked his coworkers to also cease cooperating. When the FDA later claimed to have "evaluated" the accumulator with negative findings, and petitioned a court to issue an injunction against his work and writings, Reich chose not to appear in court "as a defendant in matters of basic natural science". Instead, he prepared a petition, *Response To Ignorance,* and sent it to the judge, outlining the prejudice against his work, and the contemptuous and antiscientific nature of the FDA pseudo-investigation. Unfortunately, Reich's *Response* petition was ignored by the judge, who declared that the orgone energy "did not exist", and gave the FDA everything they asked for. The FDA won the case, and an injunction was issued against

Reich. All his books bearing the word "orgone" were labeled as "advertising literature" by the court. This "advertising literature", included books by Reich that are now classics, plus all of his published research journals and papers; they were ordered banned from interstate shipment, with existing copies to be burned in incinerators. The book-burnings took place on several occasions, as recently as 1960; this included books such as *Character Analysis, The Sexual Revolution, The Mass Psychology of Fascism, The Function of the Orgasm,* and *The Murder of Christ,* plus back issues of the *International Journal of Sex-Economy and Orgone Research,* and *Orgone Energy Bulletin.*

FDA agents also subsequently invaded Reich's Maine laboratory, and put accumulators to the axe. Later, one of Reich's assistants technically violated the injunction, by moving banned materials across a state line. Reich was charged with, and eventually convicted of contempt of court, and sentenced to two years in prison. He appealed the case all the way to the US Supreme Court, but lost, and was incarcerated in Lewisberg Federal Penitentiary. He died two weeks prior to his parole date, at a time when his spirits were high, while making plans to move to Switzerland with his son and new wife. By some accounts, his treatment in the prison had been abusive. After his death, prison officials refused to release a book manuscript he had been writing, and abused his body in the autopsy. His children believe he was poisoned. Such was the tone and tempo of Reich's treatment at the hands of the FDA, courts, and penal system.

Greenfield's work on the Reich legal case did not review the laboratory or medical experiments undertaken by the FDA, but did document the hostility and treachery involved in high government and medical-academic circles. After Reich's conviction and death, letters of congratulations passed between FDA officials and high-ranking members of the medical and psychiatric communities. During the entire period, not a single scientific, research, academic, or professional organization, or a single newspaper or citizen's group, made a move to assist Reich, or to take his side, even regarding the court order to ban and burn his books and journals. Only a few of his loyal friends and coworkers raised objections to his treatment, but to no avail.

The Present Review of FDA Files

First, I should like to say that my review of the FDA internal documents, which took place from 1981 to 1982, corroborated the findings of Blasband, Baker, and Greenfield, as summarized above. Additionally, the following new considerations were gleaned from original FDA documents.

The FDA document "Investigations on the Or-

gone Accumulator" by Kurt Lion of MIT (7), summarized the results of the FDA's non-medical experiments with the orgone accumulator. Lion attempted to duplicate the orgone accumulator temperature differential experiment, called the To-T test, and the accumulator GM effect. In the To-T test, the temperature inside the orgone accumulator is compared to the temperature outside of the accumulator, or within a thermally-balanced control enclosure lacking the metals that are characteristic of the orgone accumulator. This test demonstrates the fact that the accumulator will, on sunny days, maintain a higher temperature than its surroundings. Reich considered the To-T test a clear proof of the orgone radiation inside the accumulator, and a refutation of the second law of thermodynamics. The GM experiments demonstrated a link between the orgone and classically-described radioactivity; a GM tube which is adequately saturated and charged with orgone energy will yield a different count rate for the same radioactive source, or for background radiation, as compared to an uncharged GM tube. In my review of Lion's attempt to duplicate these experiments, I found a number of significant discrepancies and fatal methodological flaws, some of which were not mentioned in the previously cited studies.

For example, in Lion's version of the To-T test, there was no mention in his paper of the time of day when temperature measurements were made, or of the thermal dynamics in rooms where the experiments proceeded. There was no mention of whether the laboratory was at ambient outdoor temperature, or if it was heated. As Lion's experiments proceeded in the Massachusetts winter, we might *assume* his lab was heated, but there is no real way to tell as this vital information was neither given nor, apparently, considered important. However, in the properly performed To-T test, temperature differentials of fractions of a degree are monitored, and considered significant. It requires great care and skill in the design and performing of this test as the opening of a door, or the proximity of a window, human body, or heating system, may influence the measurements. Even so, on one of the figures given in Lion's paper, which charted the changing room temperatures (7:figure 4), appeared the note "door found open"; this note corresponded to a huge decline, the largest one, in the graphed data for room, control, and accumulator temperatures. From this notation, however, we can infer that Lion's lab was heated and not kept at ambient, outdoor temperature. This is a significant point, as Reich's accumulator experiments generally proceeded at ambient, environmental temperatures.

As noted above, the time of day of temperature measurements was not recorded, nor could one tell how many readings were made per day; Lion's graphs and figures indicated that on some days several temperature measurements were recorded, while on weekend days no measurements were taken at all. Temperatures appear to have been recorded at undetermined intervals during weekdays, with a cessation of "record keeping" on weekends when everyone went home. Artifacts in the graphed data were glaring: huge dips and peaks corresponding to thermal artifacts (opening and closing of doors and windows, heating system on and off, etc.) and long stretches of straight lines drawn in on the graphs corresponding to weekends when temperature measurements were not made. This latter point is most annoying, as decent and clean scientific method requires that missing data be identified as such, and not glossed over by drawing straight lines between data points separated by days with no data.

Lion also failed to contrast his experimental results on To-T against local meteorology. Reich, for example, found that there was a weather-dependent pulsation of the orgone charge inside the accumulator. It developed its strongest charge (and highest To-T) on clear, sunny days, and its weakest charge (and lowest To-T) on overcast, rainy days. Lion was apparently ignorant of this fact regarding accumulator functioning and, hence, looked for a constant positive thermal effect from the accumulator. At one point, it seemed possible that Lion's own data could be re-evaluated against the weather records for Massachusetts, to see if such a meteorological pendulation actually had been recorded.

While weather records for the region were available, they proved to be of little use. The FDA had either lost, or had never been given, Lion's original data tables. And the poor quality of the photocopied graphs and figures which I did receive from the FDA made detailed interpretation of his numerical data literally impossible. For instance, lines on the graphs for room, control, and accumulator temperatures crossed repeatedly could not be distinguished from each other in the greytones of the photocopy. Worse, the important numerical scales were also missing from the sides and bottom of the graphs, prohibiting any interpretation whatsoever of the absolute values or times of the data points. This is clearly a most sloppy procedure for any experiment.

There also was not enough detail in Lion's paper to rule out the possibility that his "control" device may have had steel wool in it, thereby making the control an accumulating device itself. Lion stated:

"From December 15 to December 20 we also used a control box, which we had built, with the same dimensions as the orgone energy accumulator, but without metal lining "[emphasis added: J.D.]" (7:8).

If steel wool had been included in the construction of the control, it might explain why the control readings were as high as those of the accumulator — both accumulator and control temperatures in the first of Lion's experiments were higher than room temperature (7:figure 3). In reviewing the matter I got an ill feeling that Lion's "control" was actually a small accumulator with its inner metal lining stripped away.

Lion also attempted to evaluate Reich's experimental findings regarding the GM effect of the orgone energy. In Reich's work on the GM effect, in addition to standard GM tubes, a number of special glass and metal high-vacuum tubes were assembled, which he called *vacor* tubes, and the experiments took place in a highly charged, orgone energy accumulator room, on a dry mountain top in rural, unpolluted Maine. Lion did not attempted to construct any of these special vacor tubes, or go to mountain-top locations, or build any of the powerful accumulators needed to develop the strength of orgone charge discussed by Reich for verification of the effects. There is evidence, instead, that Lion's laboratory was contaminated with orgone-affecting low-level radiation.

In his paper, Lion first discussed the calibration curve for his GM apparatus, which is a standard procedure. His calibration yielded a measurement plateau of around 70 counts per minute (cpm), which is quite high (7:figure 6); slightly lower count rates, on the order of 40 to 60 cpm, were recorded both inside and outside of the accumulator during the actual tests. This strongly suggests that the count rates came solely from "background" sources in his lab (7:figures 6 & 7). If so, it is a high level of background radiation, particularly for a location close to sea-level. For the record, I have worked in different physics labs in North America, including one adjacent to a nuclear reactor, and have made numerous GM measurements of background radiation in the field, around the environs of creaking and leaking nuclear power plants, and also during a period of nuclear fallout following a Chinese atmospheric bomb test, when children were advised not to play outdoors, and mothers and babies were cautioned not to drink milk. A sustained 70 cpm is higher than any background level I have ever measured, and it is more than *twice* what one would expect from natural background sources alone. By comparison, my GM readings 10 miles distant from the leaky Hanford Nuclear Facility, in Washington State, ranged only from 35 to 50 counts per minute. Based upon this, I conclude that Lion's laboratory was contaminated with low level radiation. Another of his calibration graphs supports this inference (7:figure 7), and suggests that Lion was unconcerned with the effects of low-level radiation; his radium-glow wristwatch alone was giving off around 700 cpm at some unspecified distance "near" the Geiger tube.

For Reich's orgone experiments, this is a crucial factor. Such a radioactive environment would have driven the orgone in the accumulators into a wild frenzy, a phenomena he had observed and wrote about, and called the *oranur effect*. In 1951, Reich published the *Oranur*

Experiment,(8) and Lion knew about this publication, and cited it.(7) However, he apparently did not fully read the work, or understand the phenomenon or believe that it was real. Based alone upon on the presence of oranur in Lion's laboratory, the results of even an otherwise accurately designed and conducted To-T test would *not* be considered valid. After discovering the oranur phenomenon, Reich studiously excluded radioactive materials, and other oranur producing substances and devices, from the environs of his accumulators.

While Lion reported "no effect" from the accumulators on radiation measurements made with his GM apparatus, in fact, his own measurements showed a 15-20% *reduction* in cpm when Geiger tubes were placed inside the accumulator. He noted this himself:

> "The outcome of this experiment did not show any significant change that had occurred to the tube or to the Orgone Energy Accumulator. The results are the following: Outside of the box measured: 1.04 and 0.99, [counts per second]Inside the box: 0.79 and 0.77. [counts per second] That means a recession of the reading of about 20 percent when the counter was inserted into the Orgone Energy Accumulator...Again, less effect *in* the accumulator than outside",(7:9-11)

While it is possible that the reduced count rate might have been due to the accumulator acting as a *shield* against stray radiation in Lion's contaminated lab, his observed reduction in count rate is also in keeping with one of the effects of the orgone accumulator, noted earlier by Reich: an initial *squelching* of count rate occurs when a GM tube is soaked in an orgone accumulator. Lion should have anticipated this initial squelching effect, but was apparently not even aware of it, and did not make measurements at frequent enough intervals to properly determine the effect of the orgone upon the tubes. He acted as if it violated Reich's theory, when in fact, it partly confirmed it.

In a later attempt to secure more legible copies of Lion's paper and data, I exchanging letters and phone calls with the FDA Freedom Of Information Unit for over a year. Finally, I received a letter from them acknowledging that they had entirely *lost* Lion's original data and report. The only copy they maintained was the illegible and sloppy one I had previously received.

> "As I stated in our telephone conversation yesterday, we are unable to locate the original charts of the study done by Dr. Lion on Dr. Reich's case. Therefore, the graphs that were previously provided to you are the best

copies available." (9)

Later, in an attempt to secure the needed data, I wrote directly to Lion at his home, only to find that he had passed away a few years before. His records on the Reich case had been thrown out by his family shortly following his death. However, his son, Dr. John Lion, then a psychiatrist at the University of Maryland School of Medicine, wrote to me on the matter:

> "It is strange that you should undertake this task. Without wishing to intrude upon your own interests, you are no doubt aware that Dr. Wilhelm Reich started out his career in psychoanalysis as a brilliant clinician but, lamentably, ended it quite deranged. His orgone accumulator appeared to represent the height of a delusional folly and *I recall quite well that my father was called upon to prove that the box was just a box and that Dr. Reich was a fraud.* Dr. Reich went to jail and the whole chapter of orgone is a tragic story." [emphasis added, J.D.](10)

Indeed, a tragic story...

Conclusions

From the above, several important points emerge:

1) We may reconstruct a picture of Kurt Lion's laboratory: wintertime, heated through an unknown means, doors and windows opened and shut at random, temperatures oscillating widely, abandoned over weekends, possibly stuffy, and mildly radioactive! I cannot think of a *less* hospitable environment in which to carry out these very sensitive accumulator temperature experiments.

2) The one study the FDA had against Reich which they held to possess some semblance of a quantitative experimental approach, namely the To-T test, failed to adhere to the most basic and well-known tenets of experimental design and method, even from the more narrow viewpoint of conventional thermodynamics. Furthermore, the original documentation of the experiment, and the important raw data, are *lost,* with the only surviving copy so illegible as to render it unintelligible.

3) Psychiatrist John Lion, the son of Kurt Lion, recalls "...quite well" that his father was "...called upon to prove that the box was just a box and that Dr. Reich was a fraud." This is quite a different matter from Lion being asked to *evaluate the accumulator,* and fully supports the assertion that the FDA investigation was designed, from the start, to perpetrate a fraud upon the courts, and "get Reich" through any means possible.

My own study of the FDA experimental protocols both confirmed and slightly extended the criticisms previously given by Blasband, Baker, and Greenfield. The FDA scientists attempting to duplicate Dr. Reich's work simply had no idea of what they were doing. Not a single one of the experiments conducted by FDA scientists to test Wilhelm Reich's theories can be construed as a proper duplication of the original experimental record. Some of the FDA scientists demonstrated what can only be called a grotesque incompetence with respect to traditional experimental design and laboratory technique. Further, a very negative bias and contempt for the work was revealed by their failure to take note of positive experimental evidence when it occurred, and by their failure to incorporate crucial experimental controls, the necessity of which would have been apparent from even a cursory reading of the literature then available on orgone biophysics.(11)

REFERENCES:

1. Blasband, R.A.: "An Analysis of the Food and Drug Administration's Scientific Evidence Against Wilhelm Reich, Part I: The Biomedical Evidence", *Journal of Orgonomy*, VI:207-222, 1972.
2. *Rosenblum, C.F.: "An Analysis of the Food and Drug Administration's Scientific Evidence Against Wilhelm Reich, Part II: Physical Concepts", *Journal of Orgonomy*, VI:222-231, 1972.
3. *Rosenblum, C.F.: "An Analysis of the Food and Drug Administration's Scientific Evidence Against Wilhelm Reich, Part III: Physical Evidence", Journal of Orgonomy, VII:234-245, 1973.
4. Greenfield, J: "Wilhelm Reich Vs. the USA", W.W. Norton, NY, 1974.
5. Wolfe, T.: "Emotional Plague Versus Orgone Biophysics", Orgone Institute Press, NY 1947.
6. Sharaf, M.: *Fury on Earth, A Biography of Wilhelm Reich*, St. Martin's/Marek, NY, 1983.
7. Lion, K.: "Investigations on the Orgone Accumulator", Food and Drug Administration Document, from Kurt Lion of MIT to Robert Heller of the FDA, 16 April 1953.
8. Reich, W.: *The Oranur Experiment: First Report (1947-1951)*, Wilhelm Reich Foundation, Maine, 1951; Originally appeared as *Orgone Energy Bulletin*, III(4):185-344, 1951.
9. Letter to J. DeMeo from W. Crawley, FDA F.O.I. Staff, 11 May 1982.
10. Letter to J. DeMeo from John Lion, 18 January 1982.
11. DeMeo, J.: *Bibliography on Orgone Biophysics, 1934-1986*, Natural Energy Works, 1986.

* A pseudonym for Dr. C.F. Baker, maintained while he was a medical student.

Emotional Plague Report:

❖   It has come to our attention that someone appeared on the David Letterman television show, in the role of "quack debunker", wherein an attack on Reich and the orgone energy accumulator was made. Reich was apparently called the "greatest charlatan of all times", who "made a lot of money" selling the accumulator. An actual orgone accumulator and shooter, along with other supposedly "quack" items, were displayed. If anyone knows the individual who made this attack, please send the details to the Editor. This section of the *Pulse* will be reserved for discussion or announcements of these kinds of attacks, and what might be done to counter them. In future issues, we hope to present case studies of EP attacks against life-positive works, wherever they may occur.

❖   The Laboratory recently obtained a booklet announcing an August 1987 Federal Court ruling against the American Medical Association, the American College of Surgeons, and the American College of Radiologists. All three were found guilty of "Conspiracy" to destroy the profession of Chiropractic medicine. The pamphlet gives a brief history of other court rulings against the A.M.A., and the kinds of anti-health activities it, and other major medical associations, have undertaken over the years. Those familiar with the jailing of Reich, and the banning and burning of his books or the equally reprehensible attacks against other medical pioneers, will not be surprised by the facts revealed in this pamphlet. It is available from the Motion Palpation Institute, 21541 Surveyor Circle, Huntington Beach, CA 92646 (714) 960-6577.

❖   The film "WR Mysteries of the Organism" continues to make its rounds, distorting Reich's work. Most recently, it was shown at the U.C. theatre in Berkeley, California. Many people who see this film, and who know little of Reich's work, come away with a most horrible false impression, of Reich as a sexual freedom-peddler, or pornographer. In some cases, interested individuals have effectively distributed flyers to movie patrons warning them that the film does not accurately portray Reich or his views, and giving them a few citations of his writings for future reference.

Climate Features and Unusual Phenomena

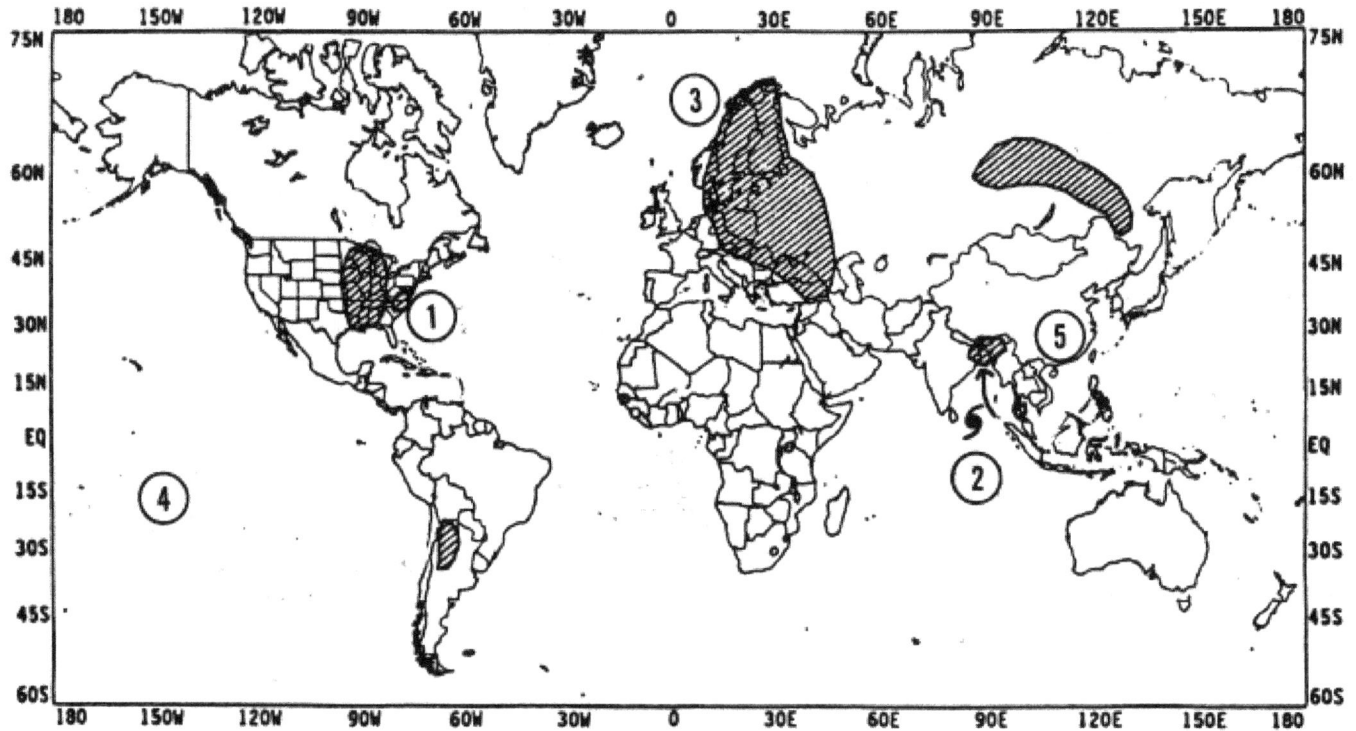

FOR THE WEEK ENDING DECEMBER 3, 1988

Persistent Conditions (shaded)
ARGENTINA: Below normal precipitation persists (23 weeks).
EAST ASIA: Warm conditions remain (8 weeks).
EAST EUROPE: Unusually low temperatures prevail (6 weeks).
CENTRAL USA: Unusual wet conditions continue (5 weeks).
THAILAND: Heavy rains reported (2 weeks).
IRELAND: Air pollution crisis lasts 6 days, from coal fire smokes; sunlight greatly diminished.

Transient Events (numbered)
(1) Nov. 29: N. CAROLINA & VIRGINIA: Tornadoes strike area; several killed, 100+ injured.
(2) Nov. 29: TROPICAL CYCLONE, FLOODS, TIDAL WAVE: BANGLADESH & E. INDIA: 1000 dead, many more missing.
(3) Nov. 29: SCANDINAVIA: Violent snowstorm with hurricane-force winds, much damage, heavy snows, and frigid temperatures.
(4) Nov. 30: ATOMIC BOMB: France, at Fangatuafa, 17:54 GMT.
(5) Nov. 30: EARTHQUAKE: S. China, 6.7 magnitude (unconfirmed, time unknown); 600+ homes destroyed.

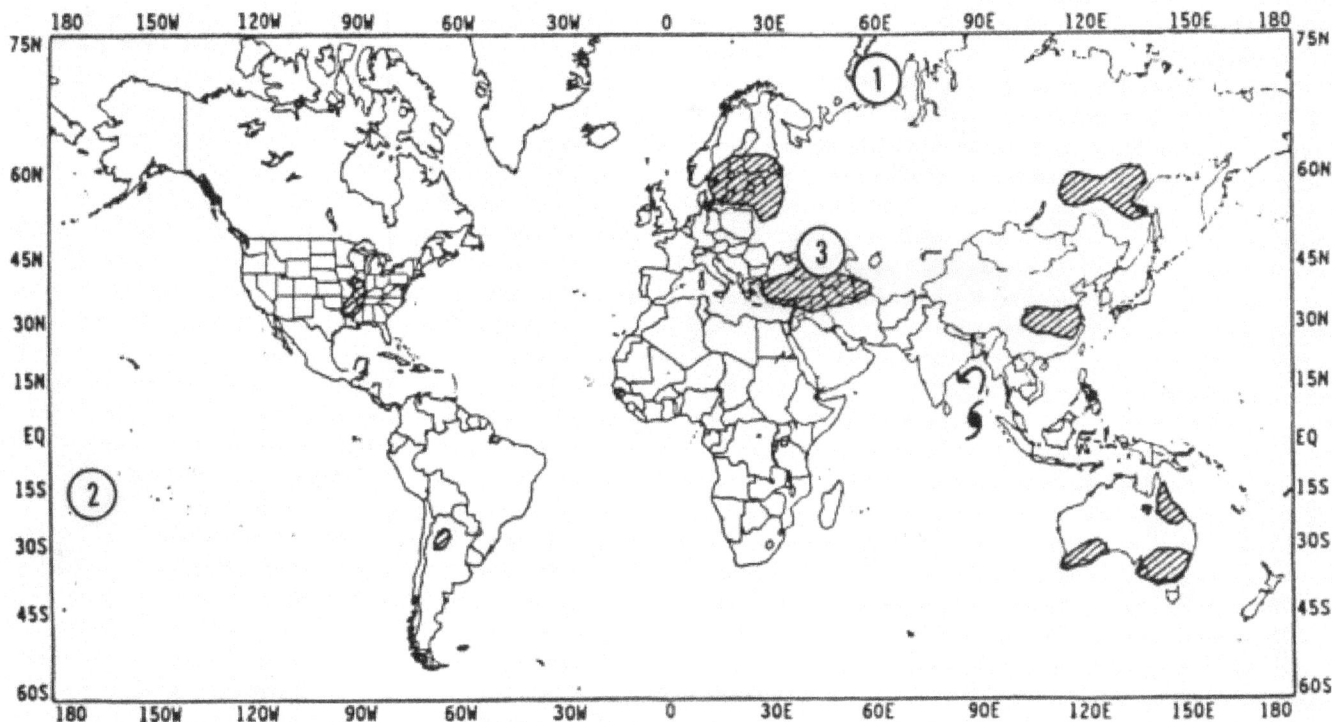

FOR THE WEEK ENDING DECEMBER 10, 1988

Persistent Conditions (shaded)
ARGENTINA: Dryness persists (24 weeks).
EAST CENTRAL CHINA: Region very dry (11 weeks).
USSR, SE SIBERIA: Mild conditions remain (9 weeks).
AUSTRALIA: Sections of country unusually wet (6 weeks).
CENTRAL USA: Unusually wet conditions diminish (Ended at 5 weeks).
USSR, MOSCOW REGION: Coldest autumn in 100 years.
COLUMBIA: Wettest rainy season in 20 years; flooding.
GABON, AFRICA: Floods
EAST EUROPE: Temperatures moderate following cold wave.
LOCUSTS: Swarms invade S. Turkey after devastating N. Africa and the Mid-East.

Transient Events (numbered)
(1) Dec. 4: ATOMIC BOMB: USSR, Nova Zemalya , in Arctic, 5:19 GMT
(2) Dec. 5: EARTHQUAKE: Tonga Islands, 6.3 magnitude, 16:05 GMT.
(?) Dec. 7: ATOMIC BOMB: Unconfirmed report of test in Central Asia.
(3) Dec. 7: EARTHQUAKE: USSR, Armenia, 6.8 magnitude, 7:41 GMT, massive, widespread damage,
      5.8 magnitude aftershock, 25,000 killed.

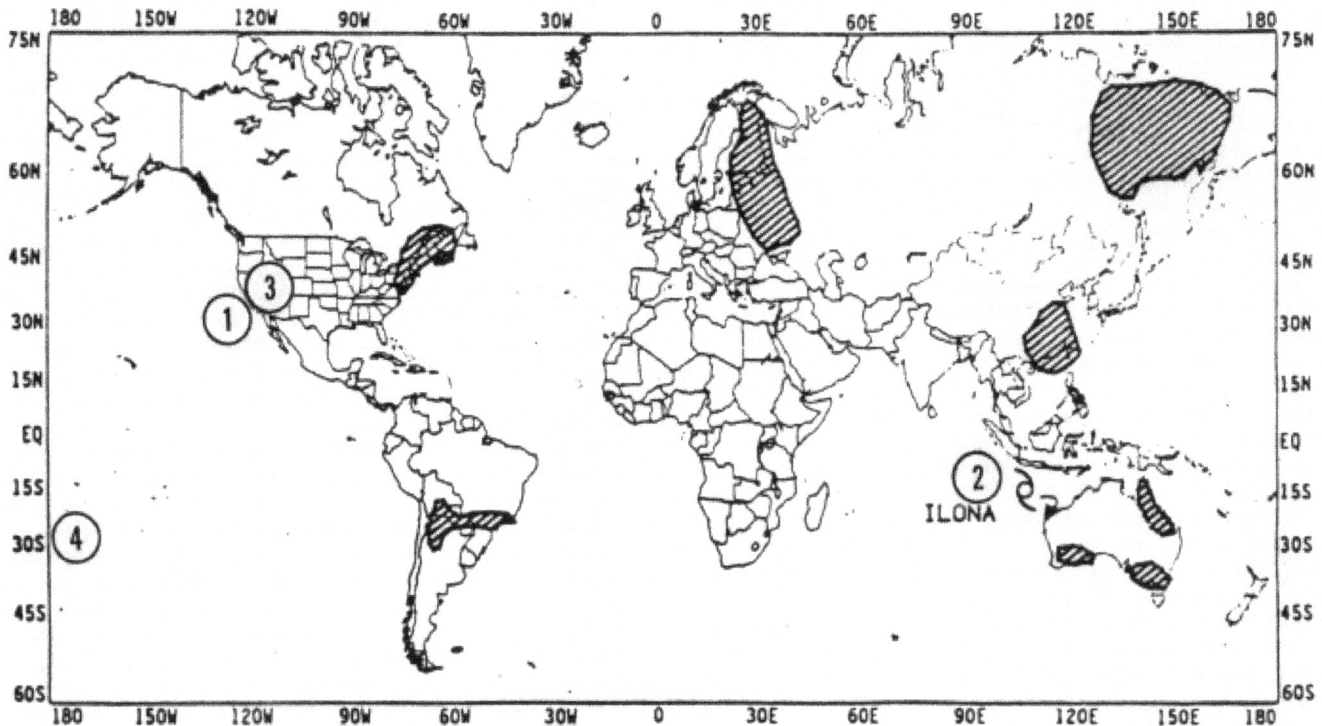

FOR THE WEEK ENDING DECEMBER 17, 1988

Persistent Conditions (shaded)
ARGENTINA, BOLIVIA, PARAGUAY, BRAZIL: Dryness persists (25 weeks).
TAIWAN, EAST CENTRAL CHINA: Region very dry (12 weeks).
EAST SIBERIA: Mild conditions remain (10 weeks).
EAST EUROPE: Low temperatures persist (8 weeks).
AUSTRALIA: Wetness persists in northeast, diminishes in south (7 weeks).
NE USA & SE CANADA: Very cold weather prevails, much snow (1 week).
EARTHQUAKES: USSR, Aftershocks continue in Soviet Armenia.
POLLUTION CRISIS: MEXICO CITY: Air pollution closes schools, disrupts normal activities.
POLLUTION CRISIS: USSR: Sturgeon in Volga River and Caspian Sea threatened by water pollution.
POLLUTION CRISIS: IRELAND: Alcoholic air pollution from brewery makes pigeons fallen-down drunk.
DEFORESTATION: THAILAND: Tree cutting forbidden in south following major losses due to erosion.
LOCUSTS: Swarms invade Canary Islands from the western Sahara; eastern Uganda also invaded.

Transient Events (numbered)
(1) FREAK WINDS: USA, SW CALIFORNIA: 100 mph winds topple radio tower, sink boats, 2 killed.
(2) TROPICAL STORM: NW AUSTRALIA: Develops and intensifies off coast.
(3) Dec. 10: ATOMIC BOMB: USA, Nevada, 19:30 GMT.
(4) Dec. 16: EARTHQUAKE: CARMADEK ISLANDS, 6.3 magnitude, 9:57 GMT.

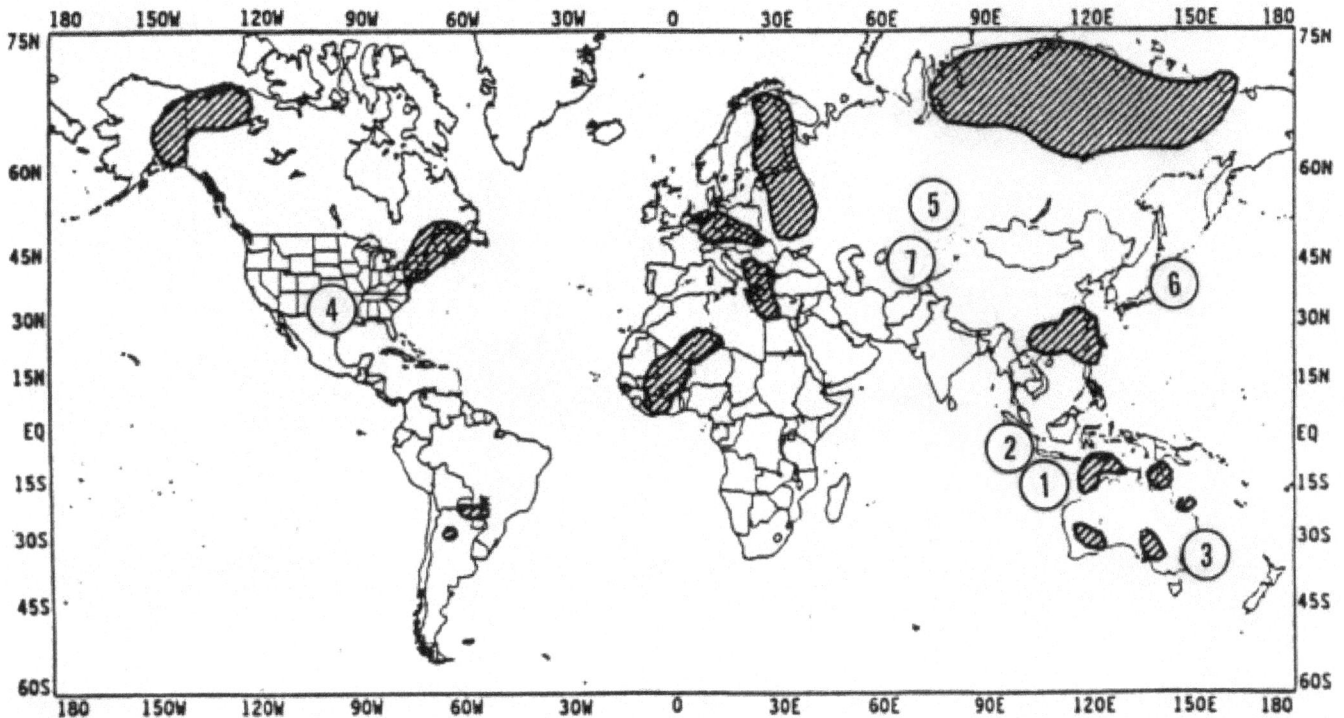

FOR THE WEEK ENDING DECEMBER 24, 1988

Persistent Conditions (shaded)
ARGENTINA, PARAGUAY, BRAZIL: Dryness diminishes (26 weeks).
TAIWAN & EAST CHINA: Dry conditions persist (13 weeks).
USSR, SIBERIA: Mild conditions linger (11 weeks).
AUSTRALIA: Wet conditions continue (8 weeks).
EAST EUROPE, NORTH & WEST AFRICA: Temperatures moderate in north and east Europe (8 weeks),
        as cold air, snow and rains moved south over Italy, Greece, and into Africa (2 Weeks).
EAST ALASKA, NW CANADA: Abnormally mild weather occurs (2 weeks).
NW USA & CANADA: Frigid conditions abate (end at 2 weeks).

Transient Events (numbered)
(1) TROPICAL CYCLONE: NW AUSTRALIA: Hurricane-force winds.
(2) FLOODS: INDONESIA: 53 die on Java; 1000 homes, 5 bridges swept away.
(3) OIL SPILL: AUSTRALIA: Giant oil slick of unknown origins washed ashore on the beaches of Sydney.
(4) BIRD DEATHS: USA: Geese in Texas die in large numbers from avian cholera.
(5) Dec. 17: ATOMIC BOMB: USSR, Central Asia, 4:18 GMT.
(6) Dec.19: VOLCANO ERUPTS: JAPAN: Mt. Tokachi, dormant for 26 years.
(7) Dec. 21: EARTHQUAKE: USSR, Central Asia, 6.0 magnitude.

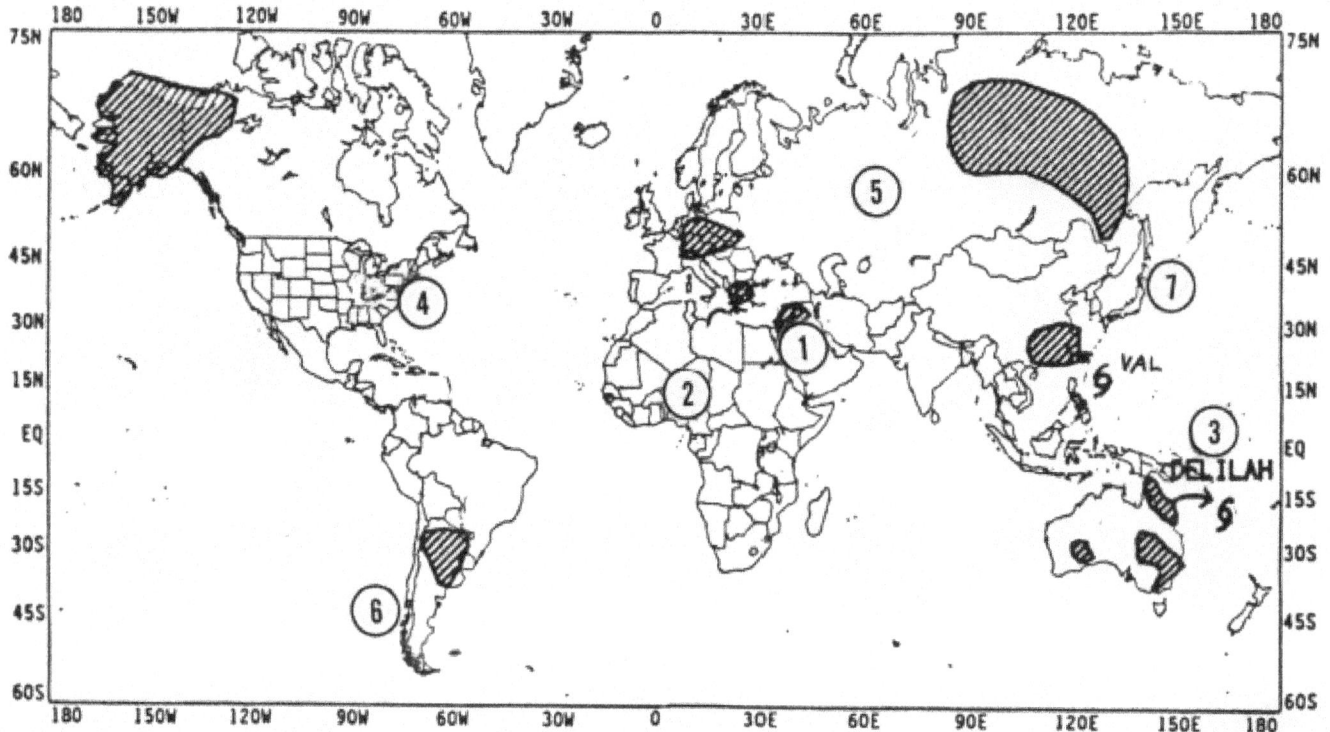

FOR THE WEEK ENDING DECEMBER 31, 1988

Persistent Conditions (shaded)
ARGENTINA: Dryness continues (27 weeks).
TAIWAN & EAST CHINA: Rains bring relief, but interior China remains dry. Severe water shortage in many
      cities, worst in 100 years; crops threatened (14 weeks).
USSR, SIBERIA: Mild weather continues (12 weeks).
AUSTRALIA: Unusual wet conditions continue (9 weeks).
CENTRAL EUROPE: Wetness diminishes (end at 5 weeks).
GREECE: Cold spell continues (3 weeks).
NW CANADA & EAST ALASKA: Abnormally mild weather persists (3 weeks).

Transient Events (numbered)
(1) FREAK SNOWSTORM, RAINS: MIDDLE EAST: Up to 12" of snow in Jordan, Israel, Lebanon; heavy
      rains in Oman, Egypt, Turkey; ten foot snowdrifts in Turkish highlands close roads.
(2) DUST STORM: NIGERIA: Harmatan winds and red dust from the Sahara disrupt activity.
(3) TROPICAL STORMS: In western Pacific, near Philippines and New Zealand.
(4) Dec.24: WINTER TORNADO, USA: Nashville, Tennessee, 1 killed, 15 injured.
(5) Dec.28: ATOMIC BOMB: USSR, Central Asia, 5:28 GMT; Mayor of Hiroshima sends message to
      Moscow urging and end to nuclear testing.
(6) Dec.28: VOLCANO ERUPTS: CHILE, Lonquimay volcano, smoke, toxic ash, lava; 1st in 100 years.
(7)VOLCANO ERUPTS AGAIN: JAPAN, Mt. Tokachi, 2nd & 3rd eruptions, mudslides.

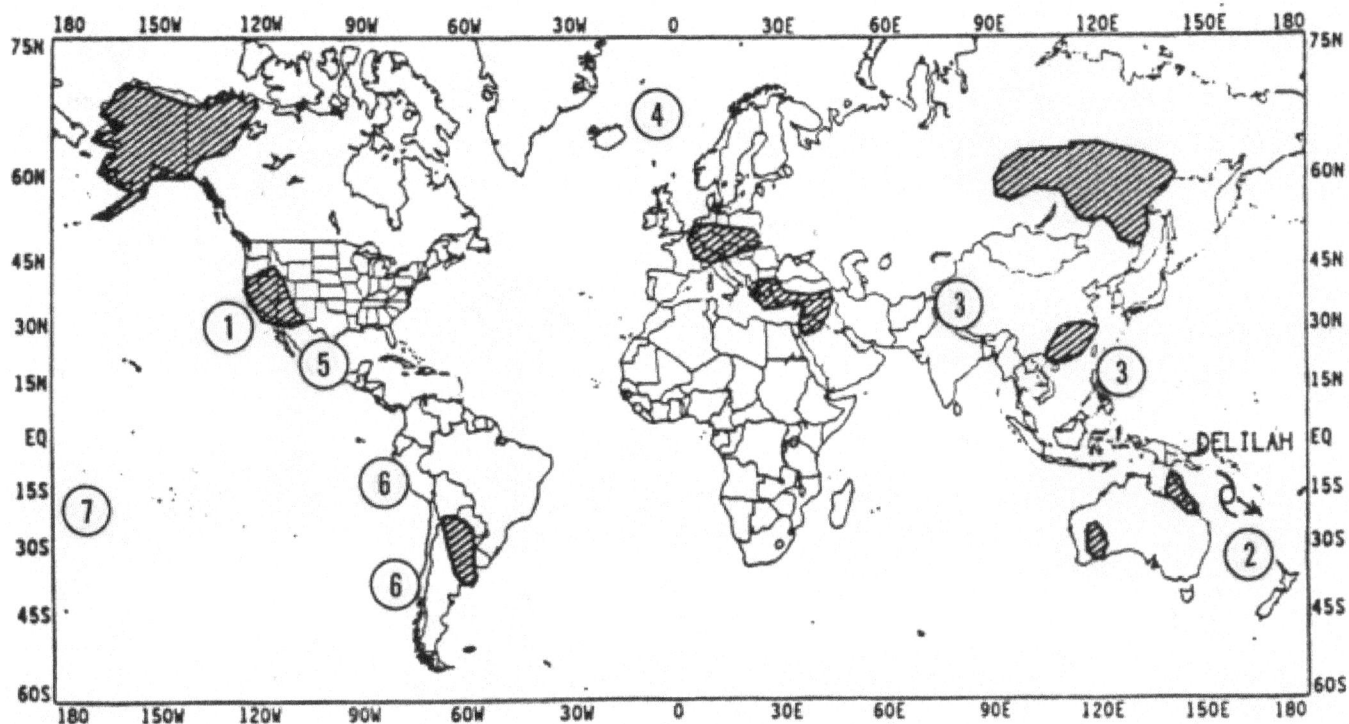

FOR THE WEEK ENDING JANUARY 7 1989

Persistent Conditions (shaded)
ARGENTINA: Dry conditions (28 weeks), and very warm conditions (7 weeks) prevail.
EAST CHINA: Rains bring additional relief from drought (end at 14 weeks).
USSR, SIBERIA: Mild conditions linger (13 weeks).
AUSTRALIA: Wetness diminishes (10 weeks).
CENTRAL EUROPE: Wet conditions end (end at 5 weeks).
MIDDLE EAST, TURKEY, GREECE: Cold spell continues (4 weeks); rare snow in Athens
NW CANADA & ALASKA: Abnormally mild weather persists (4 weeks).

Transient Events (numbered)
(1) COLD WAVE: SW USA: Unusual cold conditions, rare snowfall in Los Angeles.
(2) TYPHOON: Near New Zealand, 120 mph winds, disrupts activities.
(3) RARE COLD WAVES: Northern India, Philippines.
(4) STRONG ATLANTIC STORM: Brings rare warmth to Iceland, threatens shipping with high winds; heavy
        seas on Norwegian coast destroys power plant.
(5) POLLUTION CRISIS: MEXICO CITY: Residents warned to stay indoors air pollution at deadly levels.
(6) VOLCANO AWAKENS: PERU: Mt. Tutupaca, dormant for 100 years, now issuing smoke and gas;
        eruptions continue from nearby Mt. Lonquimay in Chile.
(7) Jan. 2: EARTHQUAKE: TONGA ISLANDS, 6.0 magnitude, 1:52 GMT.

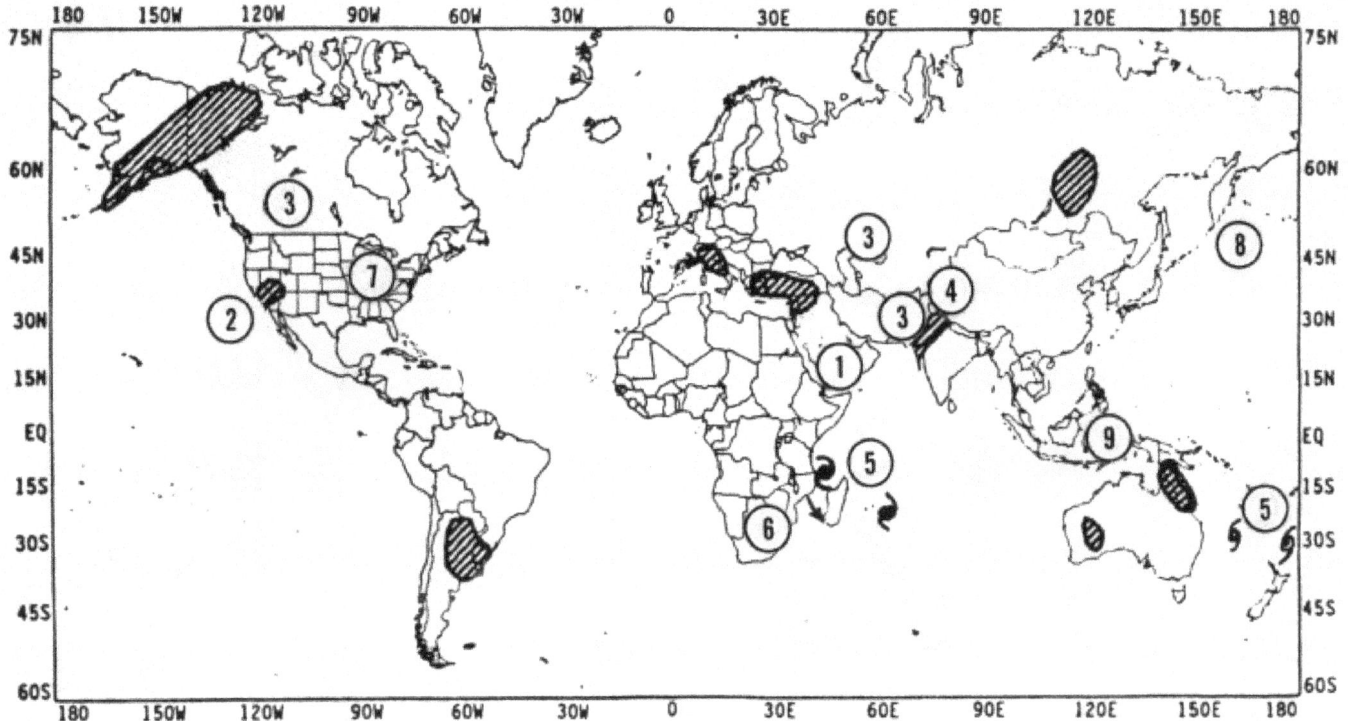

FOR THE WEEK ENDING JANUARY 14, 1989

Persistent Conditions (shaded)
ARGENTINA, URUGUAY: Very dry (29 weeks), and very warm (8 weeks).
USSR, SOUTH CENTRAL SIBERIA: Mild conditions continue (14 weeks).
AUSTRALIA: Wet conditions end (end at 10 weeks).
ITALY: Dryness develops (8 weeks).
ALASKA & NW CANADA: Abnormally mild weather continues (5 weeks).
GREECE, TURKEY, MIDDLE EAST: Cold spell continues (5 weeks).
SW USA: Abnormal Cold (4 weeks).

Transient Events (numbered)
(1) LOCUSTS: ARABIA: Swarms threaten the area from Africa.
(2) HIGH WINDS: SW USA, Los Angeles.
(3) COLD WAVES: NW India, Canadian prairies, Soviet Georgia, Cyprus, Lebanon.
(4) AVALANCHE: North India, from earth tremor; 5 dead.
(5) TROPICAL STORMS: Generate over Indian Ocean, SW Pacific
(6) SEVERE THUNDERSTORMS: SOUTH AFRICA: Crop damage.
(7) Jan.7-8: TORNADOES: USA: Illinois, Indiana, Kentucky, 50 injured.
(8) Jan. 9: EARTHQUAKE: KURIL ISLANDS, 6.4 magnitude, 13:42 GMT.
(9) Jan.10: EARTHQUAKE: ASIA, SERAM, 6.5 magnitude, 5:54 GMT.

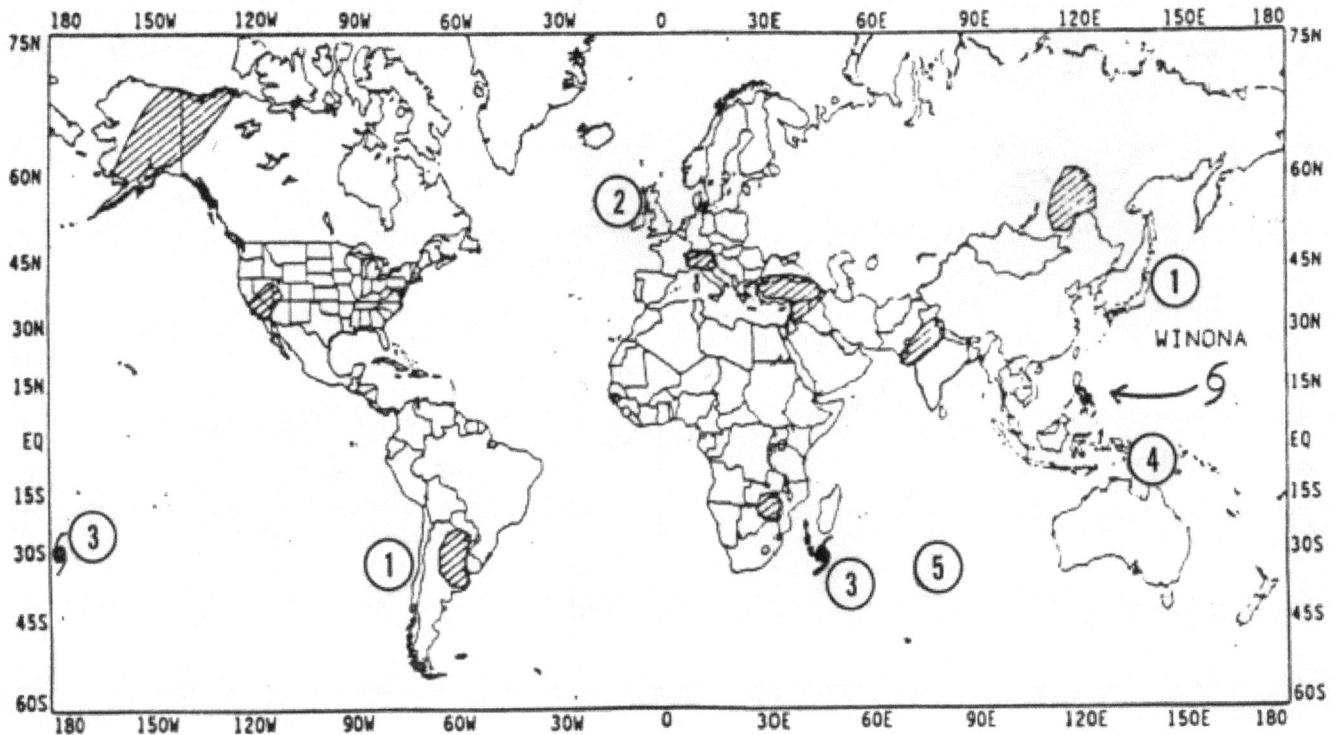

FOR THE WEEK ENDING JANUARY 21, 1989

Persistent Conditions (shaded)
ARGENTINA & URUGUAY: Abnormal dry conditions continue (30 weeks), while temperatures declined to
      near normal (end at 8 weeks).
USSR, SOUTH CENTRAL SIBERIA: Mild conditions continue; record high temps. in Moscow (15 weeks).
ITALY: Dryness continues (9 weeks).
ZIMBABWE: Dry conditions develop (7 weeks).
GREECE, TURKEY, MIDDLE EAST: Temperatures moderate (end at 6 weeks).
ALASKA & NW CANADA: Abnormally mild weather ends with onset of bitter cold (end at 5 weeks).
SW USA: Cold conditions ease up (end at 4 weeks).
NW INDIA: Unusual cold spell continues, 100+ die from exposure (2 weeks), heavy snow in Nepal.
WINTER DROUGHT: FRANCE: No snow on French Alps since early December.

Transient Events (numbered)
(1) VOLCANOES AWAKEN: CHILE: Mts. Llaima, Tolhuaca, Villarrica, spew smoke; Mt.Lonquimay spews
      lava flow of 3 miles; JAPAN: Mt. Tokachi erupts again, following its December awakening.
(2) ATLANTIC STORM: SCOTLAND, IRELAND: Hurricane force winds, ships run aground, 3 dead.
(3) TROPICAL STORMS: SE AFRICA: 120 mph winds; SAMOA: heavy rains for days.
(4) Jan.17: EARTHQUAKE: NEW BRITAIN ISLANDS, 6.5 magnitude, 0:35 GMT.
(5) Jan.20: EARTHQUAKES: MID INDIAN RISE, 6.1 magnitude at 2:03 GMT, and 6.0 at 2:15 GMT.

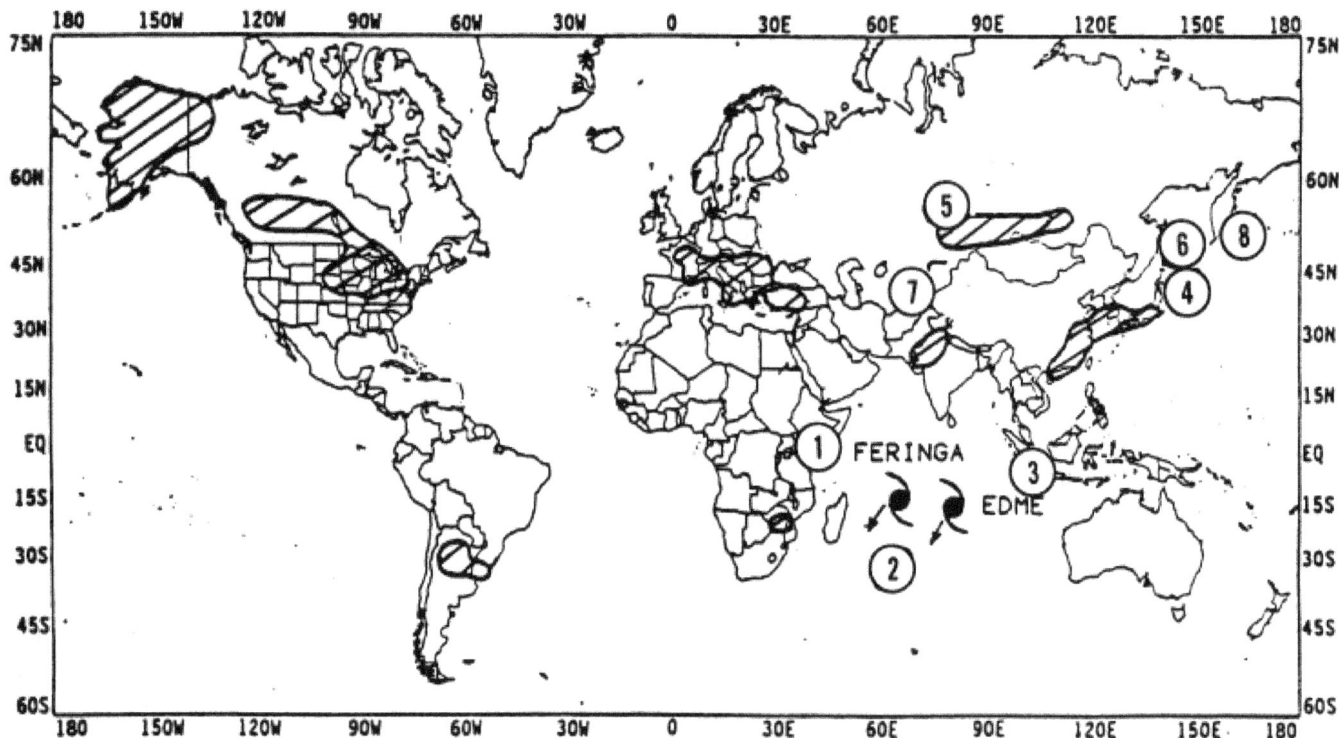

FOR THE WEEK ENDING JANUARY 28, 1989

Persistent Conditions (shaded)
ARGENTINA, URUGUAY: Area remains dry (31 weeks) and warm (9 weeks); serious crop damage and
    reduced hydroelectric capacity reported.
USSR, SOUTH CENTRAL SIBERIA: Mild conditions linger (16 weeks).
ITALY: Severe drought (10 weeks); Tiber River almost dried up, $1.5 billion in crop damage.
SOUTH EUROPE: Dryness spreads (8 weeks).
EAST CHINA, SOUTH KOREA, SW JAPAN: Abnormally wet (5 weeks).
ALASKA: Bitter cold prevails (2 weeks).
SOUTH CANADA, NORTH CENTRAL USA: Mild temperatures predominate (2 Weeks).
NW INDIA: Temperatures moderate (end at 2 weeks).

Transient Events (numbered)
(1) BLACK RAINS: KENYA: Mystery rain of dark color falls during thunderstorms.
(2) TROPICAL STORMS: Over open waters of Indian Ocean.
(3) SEVERE FLOODS: INDONESIA: A week of rain leaves 27 dead,18 missing, 12,000 homeless.
(4) VOLCANO WARNING: JAPAN: Mt. Tokachi threatens to erupt powerfully; nearby residents warned.
(5) Jan.22: ATOMIC BOMB: USSR, CENTRAL ASIA: at 3:00 GMT.
(6) Jan.22: EARTHQUAKE: JAPAN, HOKKAIDO: 6.3 magnitude at 22:20 GMT;
(7) Jan.22: EARTHQUAKE: USSR, TADZHIKSTAN: 5.5 magnitude at 23:02 GMT, extensive mudslides
    bury town of Sharora, 1000 dead.
(8) Jan.27: EARTHQUAKE: COMONDORSKI ISLANDS: 6.3 magnitude at 8:34 GMT.

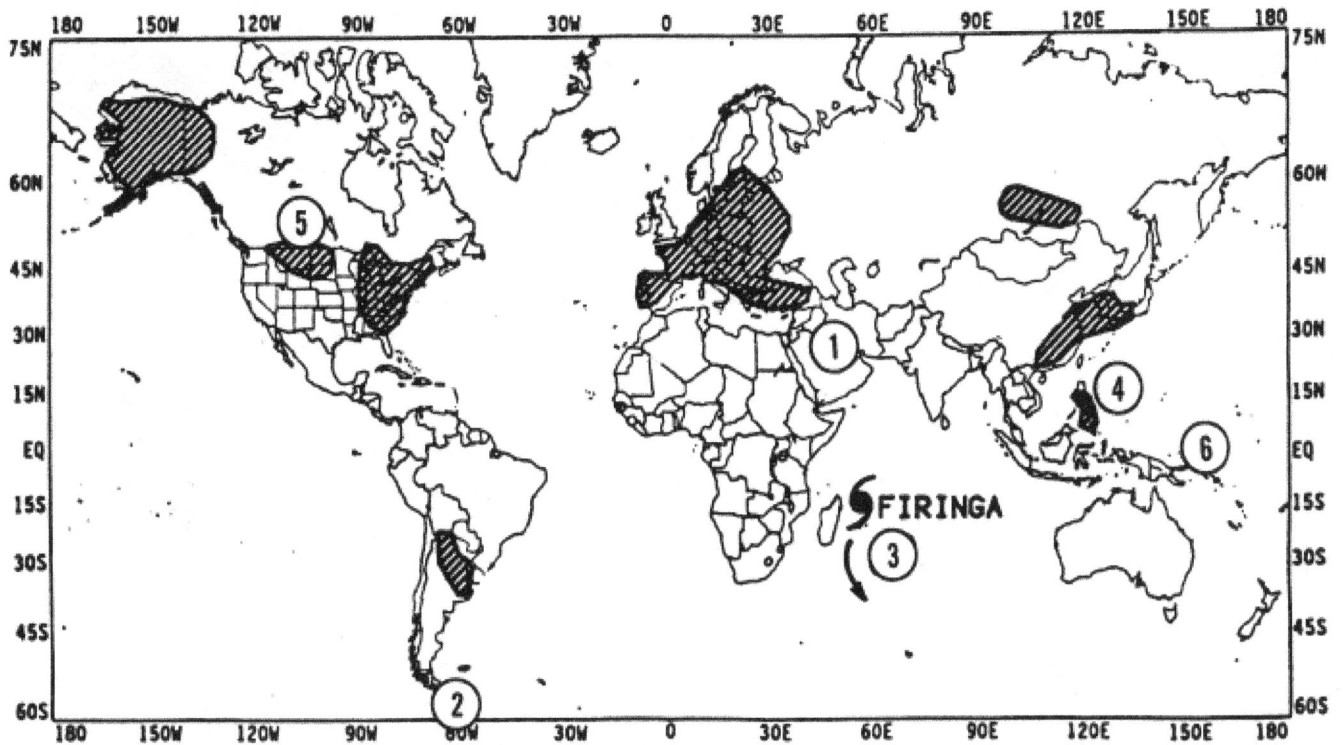

FOR THE WEEK ENDING FEBRUARY 4, 1989

Persistent Conditions (shaded)
ARGENTINA & URUGUAY: Dry conditions (32 weeks) and warm conditions (10 weeks) continue.
USSR, SOUTH CENTRAL SIBERIA: Mild conditions linger (17 weeks).
EAST CHINA, SOUTH KOREA, SOUTH JAPAN: Wet conditions diminish (end at 5 weeks).
EUROPE & MIDDLE EAST: Dryness persists (9 weeks) with above normal temperatures (4 weeks).
ALASKA: Bitter cold prevails (3 weeks); coldest temperatures (-81 degrees F) and highest barometric
    pressures (31.85 in.) ever recorded in North America.
EAST USA & SE CANADA: Dry conditions develop (4 weeks).
ITALY: Drought continues in many areas; no rain in 11 months in some areas; drinking water shortage in
    some areas; air pollution crisis worsening as atmospheric stagnation spreads.

Transient Events (numbered)
(1) ENDANGERED TURTLES: PERSIAN GULF: Hatchery given protection by the Sultan of Oman.
(2) OIL SPILL: ANTARCTICA: Seeping oil from a sunken Argentine supply ship damages penguin
    hatchery and broad areas of the coast near the US Palmer Research Station.
(3) TROPICAL STORM: INDIAN OCEAN: 125 mph winds batter island communities.
(4) FLOODS: PHILIPPINES: Heavy monsoon rains, floods, landslides last week; 120 dead.
(5) Jan.31: ARCTIC COLD BLAST: NORTH CENTRAL USA: With winds over 100 mph in Montana.
(6) Feb.4: EARTHQUAKE: NEW IRELAND: 6.1 magnitude, 22:10 GMT.

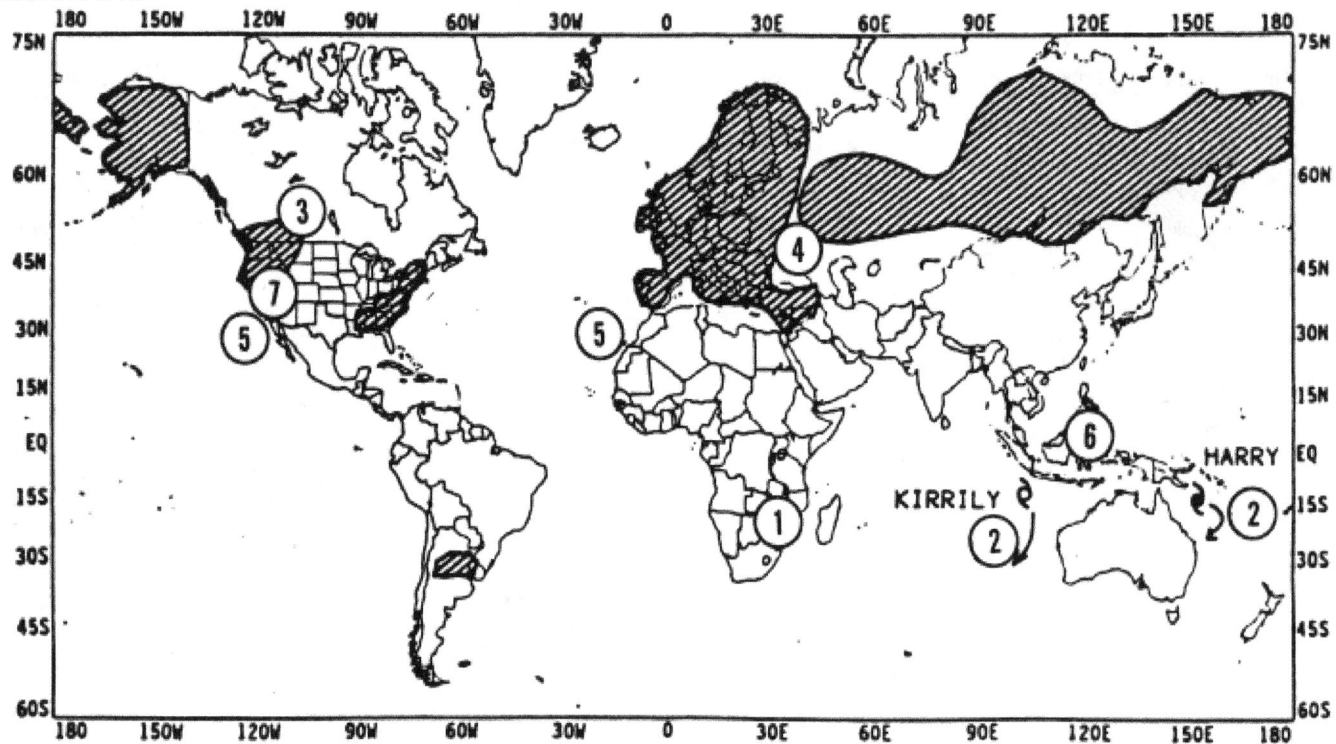

FOR THE WEEK ENDING FEBRUARY 11, 1989

Persistent Conditions (shaded)
ARGENTINA & URUGUAY: Area remains dry (33 weeks) and warm (11 weeks); record high temperatures.
USSR, SIBERIA: Mild conditions spread (18 weeks).
EUROPE & MIDDLE EAST: Dry weather persists (10 weeks); Mild in north (5 weeks).
NW USA & SW CANADA: Dryness develops (5 weeks).
EAST USA: Area still dry (5 weeks).
ALASKA: Bitter cold ends (end at 3 weeks), as cold air mass spills south into Canada.
FRANCE: Absence of normal winter snow in Alps leaves ski resorts, water planners worried.

Transient Events (numbered)
(1) FLOODS: ZAMBIA & BURUNDI:Floods & mudslides; roads, bridges out; dozens dead.
(2) TROPICAL STORMS: Near western Australia and Fiji.
(3) SWAN STARVATION: USA, IDAHO: Extreme cold threatens to starve 25% of all trumpeter swans.
(4) POISON RAIN: USSR, UKRAINE: Rainfall contaminated with thallium from a chemical plant falls on
        Chernovotsy; 150 children lose hair, suffer from irritability & hallucinations.
(5) RARE SNOWFALL: SW USA & CANARY ISLANDS: First time in 20 and 30 years, respectively.
(6) Feb.10: EARTHQUAKE: MOLUCCA PASSAGE: 6.8 magnitude, 11:15 GMT.
(7) Feb.10: ATOMIC BOMB: USA, NEVADA, 19:06 GMT.

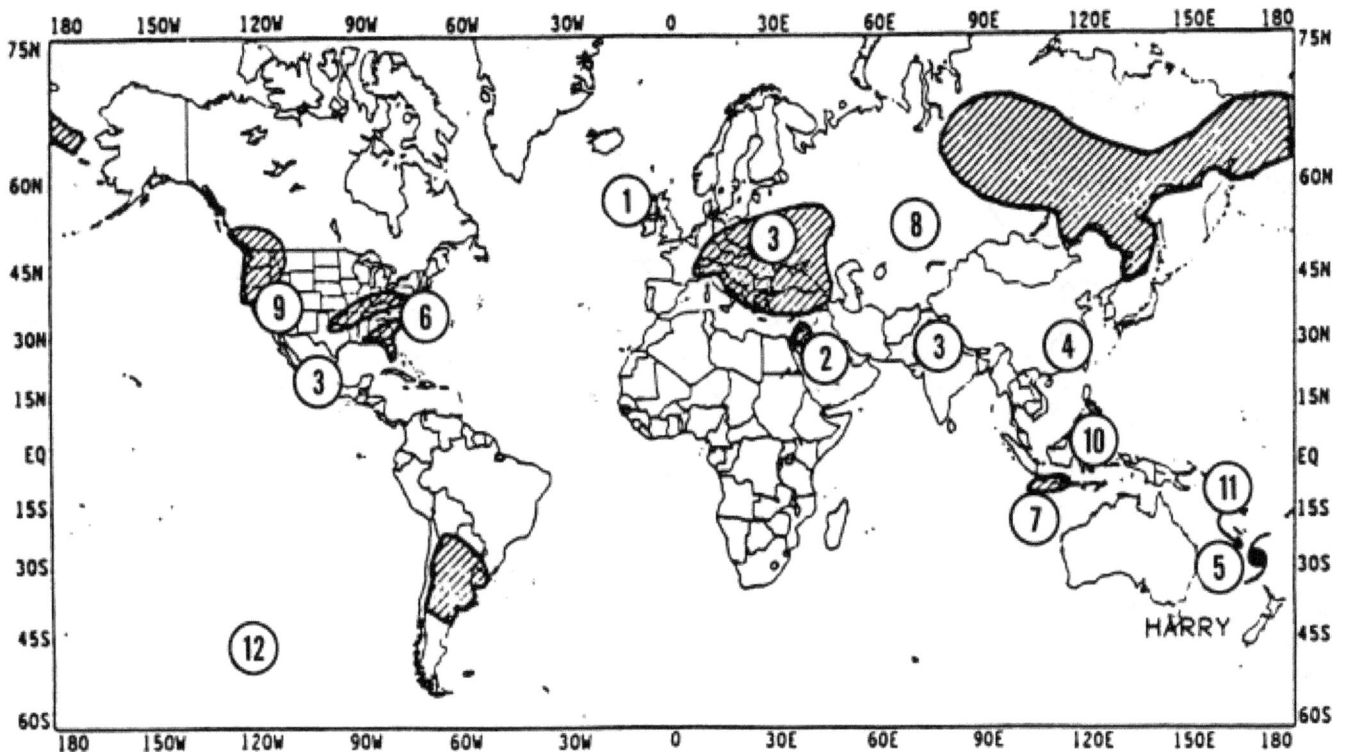

FOR THE WEEK ENDING FEBRUARY 18, 1989

Persistent Conditions (shaded)
ARGENTINA & URUGUAY: Area remains dry (34 weeks) and warm (12 weeks).
USSR, SIBERIA: Mild conditions continue (19 weeks).
EUROPE & MIDDLE EAST: Dry weather (11 weeks), with mild conditions in north (6 weeks); Italy suffers
     through its worst drought in this century; wells in many places in Yugoslavia have dried up.
NW USA & SW CANADA: Rains provide some relief along coast; dryness persists inland (6 weeks).
SE USA: Area still dry (6 weeks); Florida in water crisis.

Transient Events (numbered)
(1) FREAK STORMS AGAIN: IRELAND & NORTH BRITAIN: Two consecutive nights of hurricane force
     winds; severe, widespread damage on land and at sea.
(2) FREAK SNOW: JORDAN & LEBANON: Second major snowstorm of season; 100 cm in some areas.
(3) POLLUTION CRISIS: Severe health problems in Geneva, Switzerland, Mexico City, New Delhi, India,
     and Moscow, USSR; studies indicate damage to newborn infants.
(4) SNAKE WARNING: CHINA: Use of pit vipers as a restaurant food item threatens extinction of species.
(5) TROPICAL STORMS: EAST OF AUSTRALIA: 125 mph winds.
(6) HEAVY RAINS, SNOW: EAST & CENTRAL USA: Kentucky, Virginia, Maryland.
(7) HEAVY RAINS: INDONESIA: Up to 200 mm (7.9 inches) on Java.
(8) Feb.12: ATOMIC BOMB: USSR, CENTRAL ASIA (unannounced, time unknown).
(9) Feb.12: ATOMIC BOMB: USA, NEVADA (unconfirmed, unannounced).
(10) Feb.13: EARTHQUAKE, MOLUCCA PASSAGE: 6.8 magnitude, 11:15 GMT.
(11) Feb.14: EARTHQUAKE, SOLOMON ISLANDS: 6.5 magnitude, 6.20 GMT.
(12) Feb.16: EARTHQUAKE, EASTER ISLAND: 6.0 magnitude, 16:36 GMT.

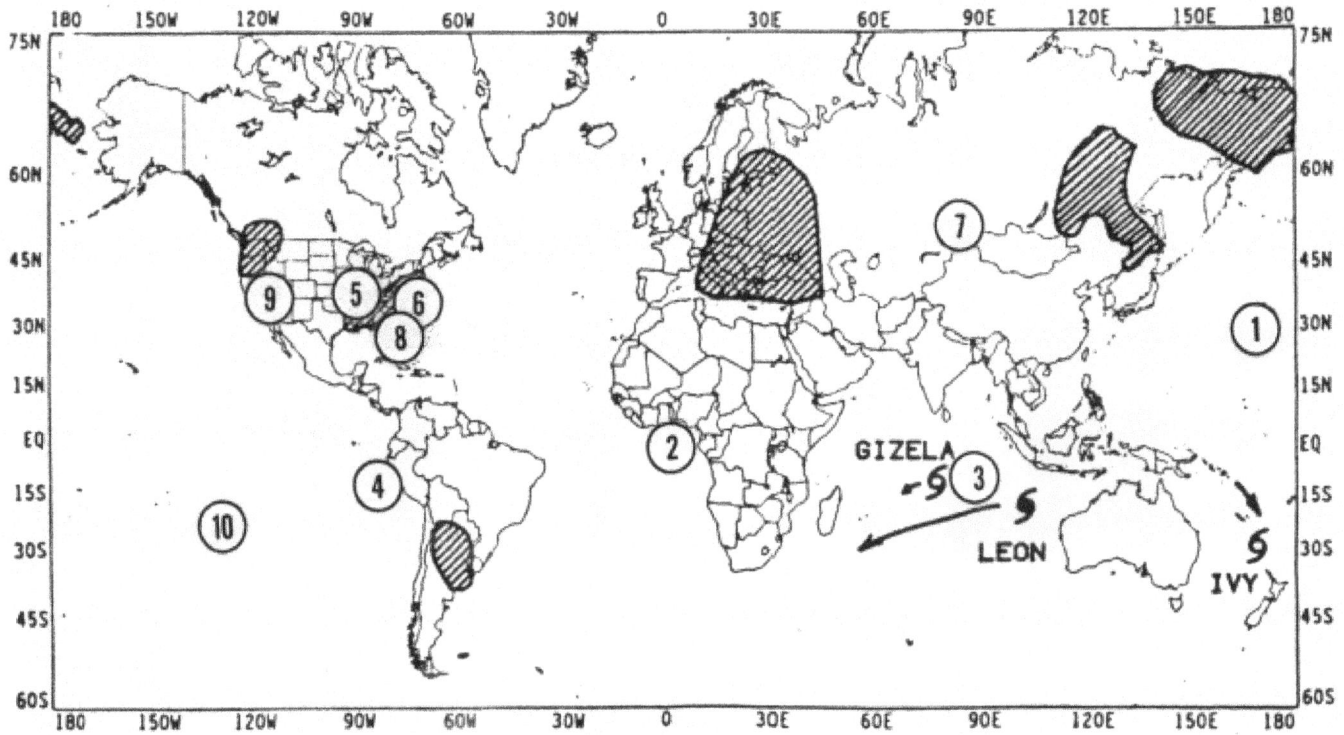

FOR THE WEEK ENDING FEBRUARY 25, 1989

Persistent Conditions (shaded)
ARGENTINA & URUGUAY: Dry conditions (35 weeks) and warm conditions (13 weeks) persist.
USSR, SIBERIA: Mild conditions persist (20 weeks).
EUROPE & MIDDLE EAST: Dry conditions (12 weeks) and mild conditions (7 weeks) persist.
NW USA & SW CANADA: Dryness returns again (7 weeks).
SE USA, FLORIDA: Dryness continues (7 weeks).
DROUGHTS LINGER OR INTENSIFY: In California & Florida, Argentina & Uruguay, Italy, Yugoslavia & Greece, Thailand, and South New Zealand.
DESERT EXPANSION CONTINUES: NIGERIA: Tree planting efforts continue, in a battle to halt the 3 mile per year southward advance of the Sahara Desert.

Transient Events (numbered)
(1) POLLUTED ECLIPSE: Lunar eclipse occurs; pollution haze blocks visibility in northerly mid-latitudes.
(2) GIANT OCEAN WAVES: NIGERIA: Coastal villages and towns swept away, people resettle inland.
(3) TROPICAL STORMS: Over the Indian Ocean.
(4) FLOODS: PERU: Torrential rains, floods, mudslides, avalanches.
(5) FLOODS: EAST USA: Ohio & Tennessee valleys.
(6) COLD WAVE, EAST USA: Snowstorms from the Carolinas to New England.
(7) Feb.17: ATOMIC BOMB: USSR, CENTRAL ASIA, (unannounced, time unknown).
(8) Feb.21: TORNADOES: EAST USA: From Louisiana to North Carolina, over 20 different events.
(9) Feb.24: ATOMIC BOMB: USA, NEVADA: 15:15 GMT.
(10) Feb.25: EARTHQUAKE: CARMADEX ISLANDS: 6.7 magnitude, 11:26 GMT.

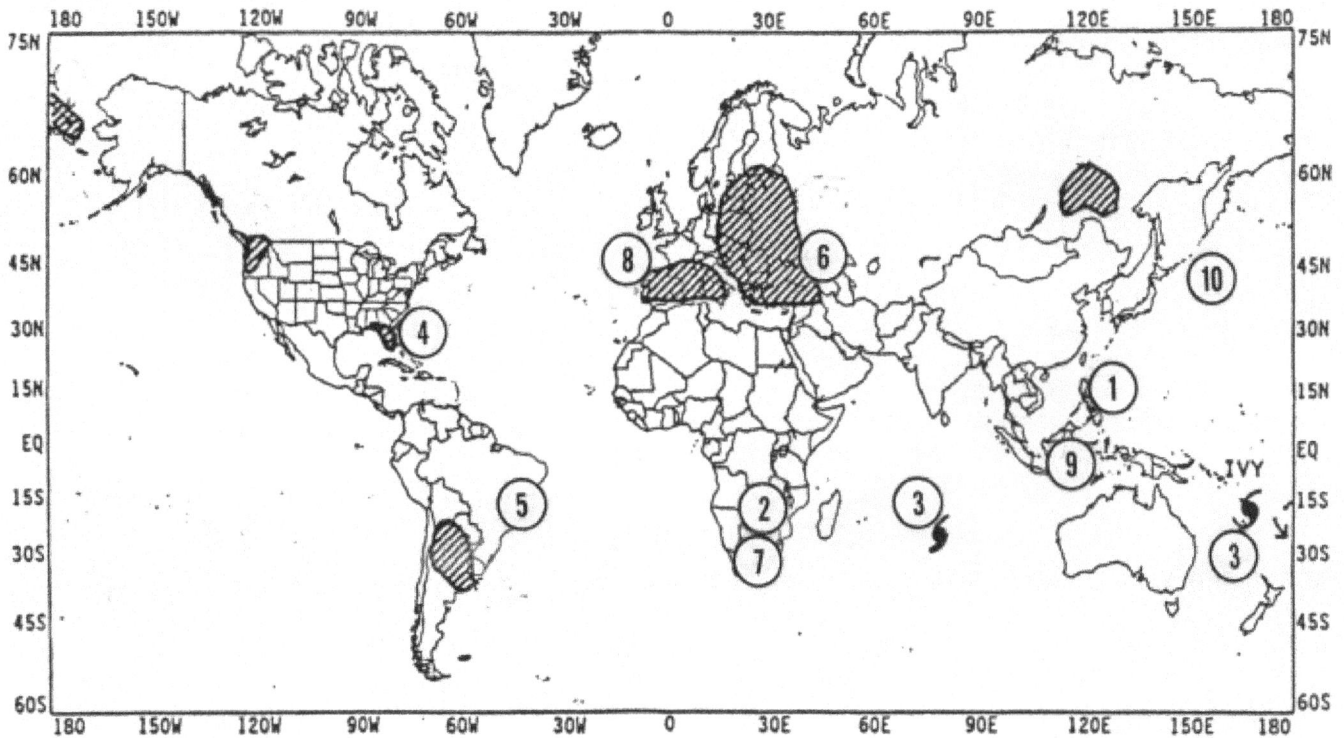

FOR THE WEEK ENDING MARCH 4, 1989

Persistent Conditions (shaded)
ARGENTINA & URUGUAY: Dry conditions (36 weeks) and warm conditions (14 weeks) persist.
USSR, SIBERIA: Mild conditions persist (21 weeks).
EAST EUROPE & MIDDLE EAST: Dry conditions ease (end at 12 weeks); but mild or warm conditions
      continue (8 weeks).
NW USA : Dryness continues (8 weeks); heavy snow in Seattle.
SE USA, FLORIDA: Rain showers east the dry spell (end at 8 weeks).

Transient Events (numbered)
(1) HOT OCEANS: PHILIPPINES: Abnormal temperatures 7 degrees above normal measured.
(2) FLOODS: SOUTHERN AFRICA: Zambia, 100 dead, 50,000 homeless, crops lost.
(3) TROPICAL STORMS: In Indian Ocean and SW Pacific.
(4) WILDFIRES: USA, FLORIDA: Everglades grassland burns more than usual due to drought.
(5) WILDFIRES: BRAZIL: Forests burning out of control in many areas.
(6) MUD VOLCANO: USSR, AZERBAIJAN: Spews mud and stones.
(7) CROCODILE INVASION: SWAZILAND: Floodwaters force migration; 10 people devoured in 3 weeks.
(8) Feb. 25-26: SEVERE STORMS: SW EUROPE: Britain, Spain, Portugal, France, Italy; 12+ killed, ships
      lost; lowest barometric pressure recorded since 1706 in the Netherlands; Snow in Britain.
(9) Feb.27 EARTHQUAKE: INDONESIA: 6.0 magnitude, 23:39 GMT.
(10) Mar.1: EARTHQUAKE: KURIL ISLANDS: 6.1 magnitude.

Atomic Bomb Tests and Earthquake Data

ATOMIC BOMB TESTS, ANNOUNCED, UNANNOUNCED, AND UNCONFIRMED

DATE	GMT	NATION/ REGION	LAT	LONG	MAGNIT.	NOTES
1988						
NOV 30	17:54	FRANCE FANGATAUFA	22.9S	138.0W	40 KT	UG
DEC.04	05:19	USSR NOVA ZEMALYA	73.4N	54.9E	35-150 KT	UG
DEC.07		USSR, CENT ASIA			<20 KT	UG unconfirmed
DEC.10	19:30	USA NEVADA	37.0N	116.3W	<150 KT	UG
DEC.17	04:18	USSR KTS	49.6N	79.6E	50-150 KT	UG
DEC.28	05:28	USSR KTS	50.0N	79.0E	0-2 KT	UG
1989						
JAN.22	03:00	USSR SEMIPALATINSK	50.0N	78.0E	20-150 KT	UG
FEB.10	19:06	USA NEVADA	37.0N	116.3W	20-150 KT	UG
FEB.12		USSR CENT ASIA				UG unannounced
FEB.12		USA NEVADA	37.0N	116.3W	20-150 KT	UG unconfirmed
FEB.17		USSR SEMIPALATINSK	50.0N	78.0E		UG unannounced
FEB.24	15:15	USA NEVADA	37.0N	116.3W	<20 KT	UG
MAR.09	13:05	USA NEVADA	37.0N	116.3W	20-150 KT	UG

MAJOR EARTHQUAKES (>M6.0), OR VERY DAMAGING EARTHQUAKES (<M6.0)

DATE	GMT	NATION/ REGION	LAT	LONG	MAGNIT.	NOTES
1988						
NOV 25:	12:48	CANADA, SE PQ,	48.0N	78.0W	6.0	Strongest in 56 years
NOV 30:		CHINA, SOUTH			6.7	600+ homes destroyed
DEC 05:	16:05	TONGA ISLANDS	15.3S	173.5W	6.3	
DEC 07:	07:41	USSR, ARMENIA	41.0N	42.0E	6.8	25.000 killed; Massive and widespread damage, M5.8 aftershock
DEC 16:	09:57	CARMADEK ISLANDS	29.7S	177.9W	6.3	
1989						
JAN 02:	01:52	TONGA ISLANDS	18.5S	174.6W	6.0	
JAN 09:	13:42	KURIL ISLANDS	46.9N	153.4E	6.4	
JAN 10:	05:54	ASIA, SERAM	3.2S	130.5E	6.5	
JAN 17:	00:35	NEW BRITAIN ISLANDS	6.1S	149.0E	6.5	
JAN 20:	02:03	MID INDIAN RISE	32.0S	79.0E	6.1	
JAN 20:	02:15	MID INDIAN RISE	41.0S	80.0E	6.0	
JAN 22:	22:20	JAPAN, HOKKAIDO	44.0N	144.0E	6.3	
JAN 22:	23:02	USSR, TADZHIK.	40.0N	68.0E	5.5	Extensive Mudslides 1,000 dead
JAN 27:	08:34	COMONDORSKI ISL.	56.0N	164.0E	6.3	
FEB 04:	22:10	NEW IRELAND	4.6S	153.0E	6.1	
FEB 10:	11:15	MOLUCCA PASSAGE	2.0N	126.0E	6.8	
FEB 13:	23:58	INDONESIA, NE	13.0S	127.4E	6.1	Halmahera
FEB 14:	06:20	SOLOMON ISLANDS	10.4S	161.3E	6.5	
FEB 16:	16:36	EASTER ISLAND COR.	56.2S	122.5W	6.0	
FEB 25:	11:26	CARMADEX ISLANDS	29.7S	177.9W	6.7	
FEB 27:	23:39	INDONESIA, NE	02.3N	128.1E	6.0	Halmahera
MAR 01:	02:42	KURIL ISLANDS	43.7N	148.9E	6.	

NOTE: Missing or unconfirmed data will be updated or corrected in subsequent issues.

Solar - Geomagnetic Data

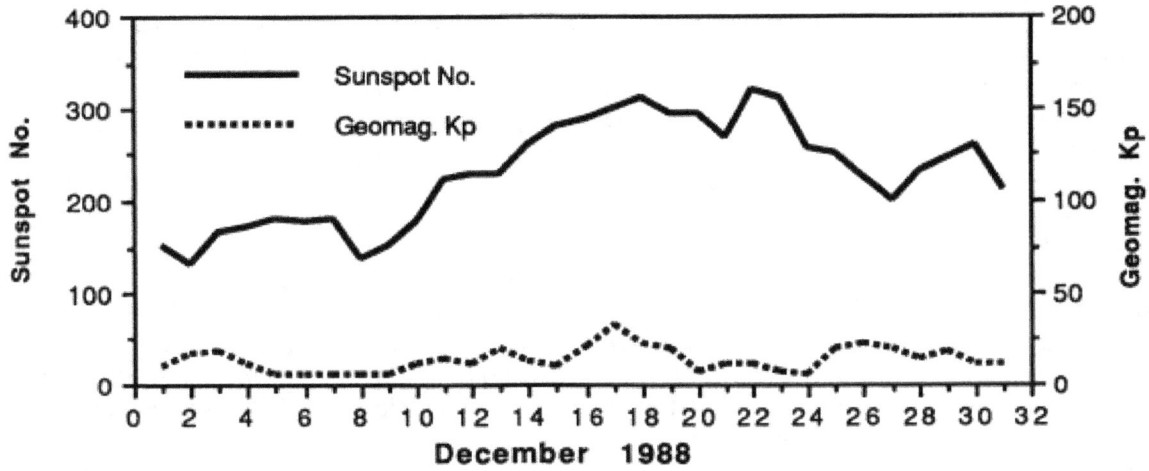

December 1988

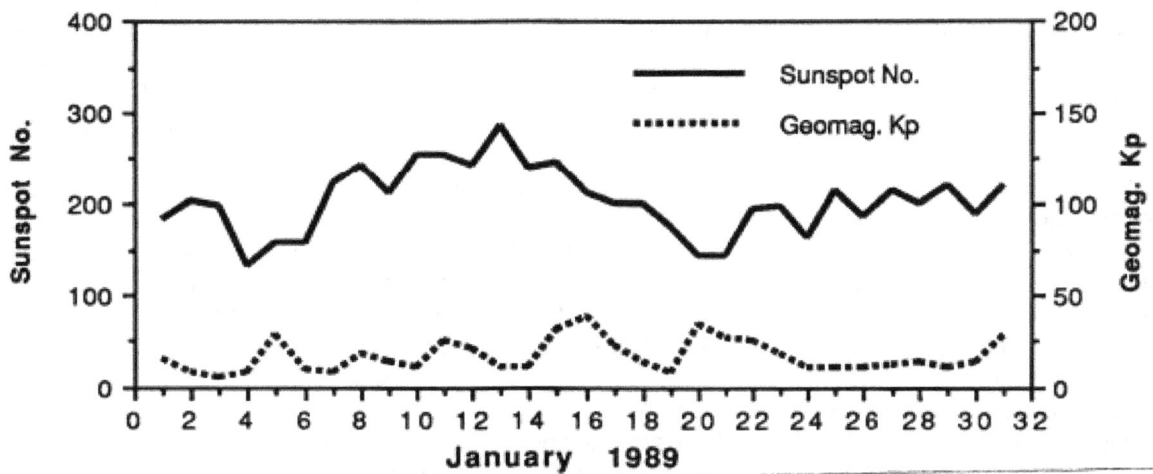

January 1989

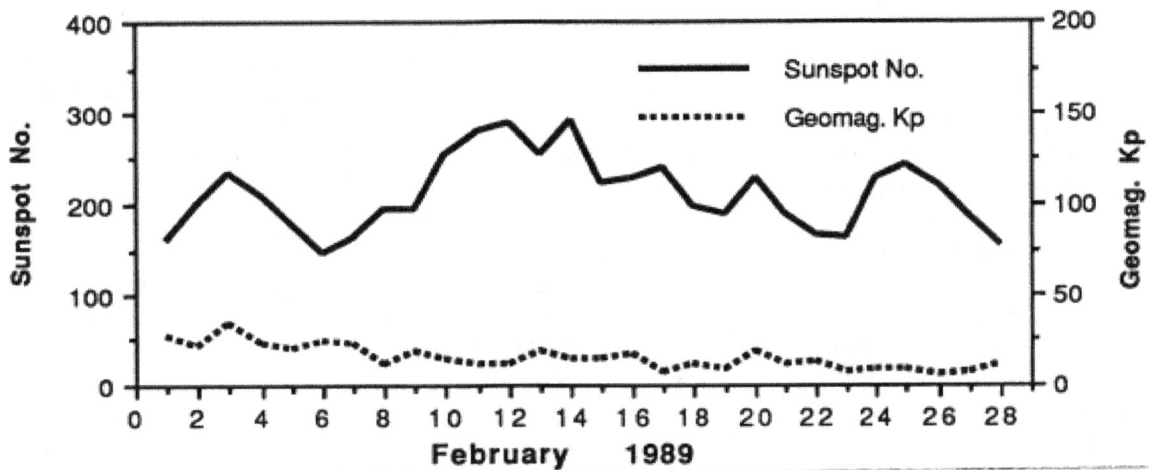

February 1989

Environmental Notes

❖ The *National Geographic* magazine recently dedicated an issue to global environmental problems. So did *Time* magazine, which ran a major story on the pollution of the oceans, and named Earth the "Planet of the Year" (as if there were others to consider). The "environment" also emerged as a political issue in the November elections. This is a welcome change from the days when "jobs", "standard of living", "economics", or "national security" were raised as objections to protecting the life on the planet. Multi-billion dollar losses resulting from major environmental catastrophes, or from centuries of environmental abuse, have convinced many about our dependence upon the larger natural environment.

❖ "Some 253 native American plants are so imperiled that they are likely to become extinct within 5 years, and another 427 will probably vanish by the end of the century". So runs a quote from *Science* magazine (16 December 1988, p.242), regarding a recent survey undertaken by the Center for Plant Conservation, which is an association of 19 botanical gardens and arboreta. More than 10% of the nations 25,000 native plant species and subspecies are at risk, largely due to habitat loss from spreading agriculture and urbanization.

❖ In addition to reports of widespread death of coral reefs in the Atlantic, comes the observation of sickness and death among sea mammals in both the Atlantic and Pacific. Europe, the USA, and other nations have used these oceans as dumps for everything from sewage sludge to radioactive waste. Environmental groups have issued warnings on these dangerous practices for years, but were ignored for the most part. Now, with sick seals and sea lions, and a threat to the vitality of ocean fish, the popular media, and even a few academic journals, have finally mentioned the possibility that there just *might* be a connection. Symptoms of the "mystery disease" in Atlantic seals includes a bionous "virus", and a weakened immune response. Congress has voted to end ocean dumping by January 1992. Eight communities in New Jersey and New York alone, for example, have been dumping 8 million tons of wet sludge into the Atlantic each year.

❖ Early in 1988, a report came over National Public Radio to the effect that naturalists in the Smokey Mountains had observed the loss of the normal bluish mountain glow about a year or two before large numbers of trees began to die. This blue glow, which is the color of the orgone energy field of the forest and mountain, has long been dismissed or explained away as an optical illusion, or purely chemical phenomenon. However, the *Pulse* recently received a clipping confirming the correlation between loss of blue orgone glow, and onset of forest death. The article, from an unidentified San Luis Obispo, California newspaper, says:

"The Smokey Mountains in East Tennessee were given the name 'Smokey' because blue mists and fog often shroud the range's rolling green landscape. Until a few years ago, those mists were the sign of a healthy ecosystem — visible vapor and fragrant resins resulting from transpiration between the Smokey's millions of trees and its atmosphere. But now National Park Service officials have begun to publicize the unhappy news that... the magical and oft-eulogized blue mists are disappearing. Shades of brown and gray have taken their place. ...the Park Service declared 1988 'Clean the Air' year because scientists finally certified that air pollution is damaging wilderness areas throughout the US. At higher elevations in the Smokies and on North Carolina's scenic Blue Ridge Parkway, nobody needs a Park Service proclamation to tell them something's wrong — trees are dying by the thousands. The red spruce trees are presently the most severely affected — their skeletons stand in stark nakedness on formerly green mountains."

❖ We have an unconfirmed report that 75% of the Kansas winter wheat crop was lost during this last winter, due to lack of snow (which melts to yield needed winter moisture). For the record, no one in Kansas attempted to contact the *Pulse* for assistance with cloudbusting, which might have been able to bring in the needed moisture and snow.

❖ A recent study made by NASA has indicated a 6% decrease in the extent of polar sea ice over the last 15 years. (*Science News*, 8 October 1988)

❖ The Monarch butterfly is a delight to all, but it will become increasingly difficult to see one. The fragile insect migrates 3000 miles twice each year, from Canada and the northern U.S.A., to a winter home in the highland forests of southern Mexico. It appears that widespread and indiscriminate logging and deforestation is wiping out their woodland habitat. (*Christian Science Monitor,* 29 Dec. 1988)

❖ On several occasions, workers in orgonomy have tried to obtain support and funding from the State of California for cloudbusting operations during drought conditions; in each case, the State declined to fund the operations. However, the California Department of Natural Resources is funding all kinds of unproven, high-risk, and environmentally-unsound ideas in its attempt to increase the state's water supply. For example, environmentalists in Plumas County recently sought and obtained a court order to halt an experimental "cloudseeding" project, wherein hundreds of gallons of liquid propane was to be sprayed directly into the atmosphere from large tanks sited on mountain tops. It is not known if the project was halted permanently, or merely delayed. (*Air/Water Pollution Report,* 9 Jan. 1989)

❖ Fallout from the Chernobyl nuclear power plant disaster is still widespread around the countryside of Byelorussia, near the burnt-out facility in Central Asia. A warning by the U.S. State Department is still in place for travelers not to eat any food grown within 50 miles of Chernobyl. Some traveler's warnings, issued by private organizations, suggest that this ban be made permanent, at least until the year 2000. And this small warning is considered to be insufficient by many health professionals. It must be remembered that each and every US nuclear power plant has the potential to have similar kinds of accidents. The much touted "containments" around our nuclear plants would be ruptured very quickly in the event of a major reactor core failure involving a steam explosion.

❖ Famine continues in the regions south of the Sahara Desert, even though the popular media does not focus much attention there anymore. In 1988 alone, 260,000 Sudanese died of starvation, with a total of 1,000,000 dead from starvation since 1983. In one Sudanese community, Abyei, 10,000 died of hunger in 1988, and virtually all children under the age of two perished. Social disturbance, political upset, and warfare has plagued the sub-Saharan region for hundreds of years, and have been exacerbated in recent years by a major population explosion, desert spreading (drought), and mismanagement of resources.

❖ Recent reports indicate a major surge of moisture into the heart of the Great Australian Desert. The moisture surge has pushed into the interior from the Pacific Ocean, westward from a point on the east coast, near Canberra. Some areas of the interior received nearly their entire yearly rainfall in just two days. Unconfirmed reports indicate the formation of huge fresh-water lakes, of never-before-seen proportions, in the desert basins of the interior. More discussion will be given on these rains in the "Climate Features and Unusual Phenomena" section of the next issue. Readers who have additional information on this event are encouraged to communicate them to the *Pulse.*

———

❖ A Family Court judge in Rhode Island recently ruled that a divorced woman with small children who invites her boyfriend to spend the night may be fined $500, or sentenced to a year in prison. The case developed during a court battle between divorced parents of three adolescent children, whose primary care was in the hands of the mother. The man sought to prevent his ex-wife from having her boyfriend sleep over. The ruling of the Family Court was recently upheld by the Rhode Island Supreme Court. (*Insight,* 3 April 1989)

❖ The November 1988 issue of the *Journal of Orgonomy* had two articles by Dr. Robert Harman, on the subject of AIDS, which should be widely distributed and read by young and old alike. One of these articles focused upon the risk of AIDS infection among heterosexuals: in short, heterosexuals who do not engage in promiscuous anal intercourse, and who do not use intravenous drugs, have a very low risk of contracting the disorder, which in large measure appears as a multi-symptom expression of a drastically weakened immune system. Dr. Harman's other article discussed the antisexual hysteria that has accompanied the limited spread of AIDS among high-risk groups. Most recently, the *Pulse* was sent some "sex-education" pamphlets that are making their way into the public schools. They are filled with untruths, wild exaggerations, and distortions about the risks of AIDS, some of which have also been spread through the popular media on television. One pamphlet, titled *Teens and AIDS, For 15 to 19 Year Olds and Their Parents and Teachers,* shows pictures of an "HIV virus" punching an "immune cell" in the nose; under the title "What is a

continued on Page 42

good way to avoid AIDS?", the pamphlet mixes reasonable advise about drugs with anti-sexual scare tactics: "Say NO to sex (abstain) until, as an adult, you are absolutely certain you have found the person with whom you want to spend the rest of your life." Another pamphlet, titled *Safe Sex for Men and Women Concerned About AIDS"*, identifies a list of "Safe Acts", which includes: "Fantasy, movies, pictures; Sex toys such as vibrators and dildoes that are not shared". Intercourse without a condom is identified, in bold red letters, as a "Dangerous Act". Nothing is said about sexual hygiene or contraception in a broader context. The message is, simply put: if you have sex, you will die; sex will kill you. Other pamphlets, supported by public money, and bearing the names of authoritative organizations (such as the Centers for Disease Control, and the AIDS Response Program), carry the same anti-sexual message. As clearly illustrated in Dr. Harman's articles, which are referenced to the most solid epidemiological evidence, young heterosexuals are not at risk for AIDS unless they engage in high risk activities, such as promiscuous anal intercourse or the sharing of hypodermic needles used in drug injections. Even promiscuous heterosexual behavior has not been proven to increase one's risk. These pamphlets do not focus upon high-risk behaviors, but do generalize the risk of AIDS from high-risk groups to the entire population, especially young people. As such, they have a hidden agenda: to frighten healthy young people who do not engage in high-risk activities, and who are at an extremely low risk for AIDS, into a very damaging sexual fear and abstinence, while simultaneously legitimizing sexual disturbances and promoting substitute gratifications for the healthier genital embrace. This kind of "sex education" is very reminiscent of George Orwell's *1984,* wherein young

couples went to regular "Antisex Meetings", sponsored by the "Ministry of Love".

❖ The *First International Symposia on Circumcision* was recently held in Anaheim, California. Presentations were given by a variety of health professionals demonstrating the significant lifelong physical and emotional damage to newborn baby boys (and girls, in some cultures) who are subject to "routine circumcision". A detailed report on this conference will appear in a subsequent issue of the Pulse. The Symposia was sponsored by the National Organization of Circumcision Information Resources Centers (NOCIRC), which is led by Marilyn Milos, R.N. Proceedings of the Symposia are planned, and can be ordered from NOCIRC at PO Box 2512, San Anselmo, CA 94960, (415) 488-9883.

❖ Immediately following the *First International Symposia on Circumcision*, the American Academy of Pediatrics cited tentative, and irrelevant biomedical evidence in a confusing new position paper which both advocated, and cautioned against circumcision of baby boys. The previous position taken by the AAP, dating

back to 1971, was clear and unambiguous, saying that "there are no valid medical indications" for routine circumcision. The newer evidence cited by the AAP consisted of a study by Major T. E. Wiswell, pediatrician at the Army Medical Center in Fort Sam Houston, Texas. Wiswell claimed that uncircumcised boys had a urinary tract infection rate 11 times higher than that of circumcised boys. However, biomedical research presented at the *International Symposium* conflicted with this claim, and suggested that improper attempts by infant caretakers to "clean" the uncircumcised baby boy's penis were largely at fault. A boy's uncircumcised foreskin is not fully retractable until around the age of 15. Many parents, however, lack this knowledge, and try to force the foreskin back to "clean" it, unnecessarily causing pain, irritation, and infection. Others at the Symposium argued that painful and mutilating surgery should not be mandated for millions of baby boys, simply because infections, treatable with simple antibiotics, occurred in a small percentage of boys. "Leave It Alone" was the advice given to parents regarding the foreskin, by the health educators and doctors at the *Symposium*.

Celestial Events Calendar

1988:

DEC	9:	New moon
	14-15:	Geminid meteor shower
	21:	Winter Solstice, 10:28 AM EST
	23:	Full moon

1989:

JAN	1:	Earth at perihelion, 5:00 PM EST
	4:	Quadrantid meteor shower
	7:	New moon
	21:	Full moon
FEB	6:	New moon
	20:	Full moon; Total eclipse of the moon
MAR	7:	New moon; Partial eclipse of the sun
	20:	Spring Equinox, 10:29 AM EST
	22:	Full moon
APR	2:	Set clocks forward for daylight time
	6:	New moon
	21:	Full moon; Lyrid meteor shower
MAY	4:	Eta Aquarid meteor shower
	5:	New moon
	20:	Full moon
JUN	3:	New moon
	19:	Full moon
	21:	Summer Solstice, 5:54 AM EDT
JUL	3:	New moon
	4:	Earth at aphelion, 8:00 AM EDT
	18:	Full moon
	30:	Delta Aquarius meteor shower
AUG	1:	New moon
	12:	Perseid meteor shower
	17:	Full moon; Total eclipse of the moon
	31:	New moon; Partial eclipse of the Sun
SEP	15:	Full moon, Harvest moon (closest to Autumnal Equinox)
	22:	Autumn Equinox, 9:21 PM EDT
	29:	New moon
OCT	14:	Full moon; Closest moon-Earth approach; strongest tides of the year
	21:	Orionid meteor shower
	29:	New moon; Set clocks back for standard time
NOV	1:	Taurid meteor shower
	13:	Full moon
	16:	Leonid meteor shower
	28:	New moon
DEC	12:	Full moon
	14:	Venus brightest in evening sky
	14:	Geminid meteor showers
	21:	Winter Solstice,
	28:	New moon

Planetary Clock

Heliocentric Planetary Configurations for the Spring Equinox, 22 March 1989

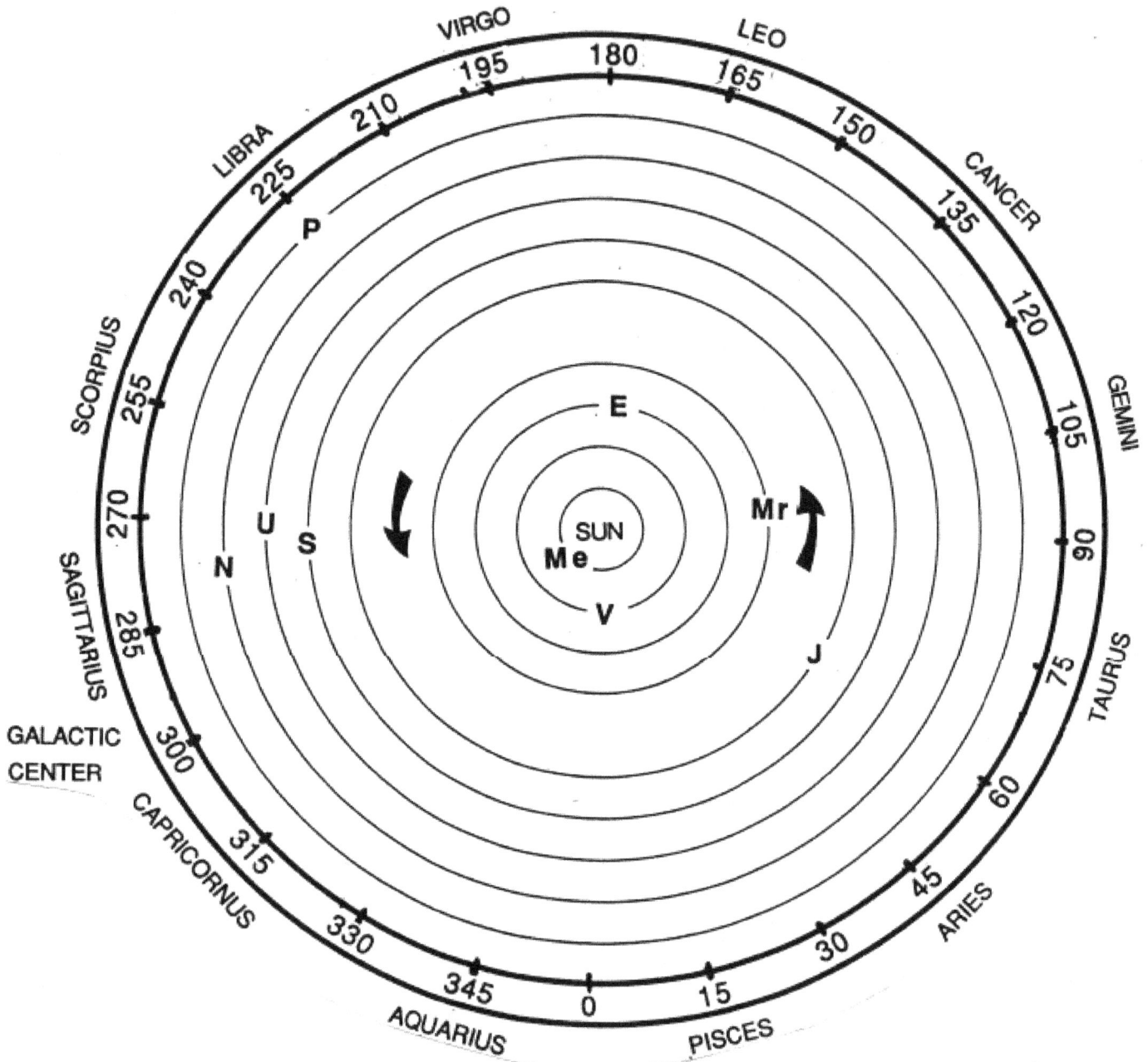

VIRGO
LEO
LIBRA
CANCER
SCORPIUS
GEMINI
SAGITTARIUS
TAURUS
GALACTIC CENTER
CAPRICORNUS
ARIES
AQUARIUS
PISCES

195 180 165 150 135 120 210 225 105 240 90 255 270 75 285 60 300 45 315 330 30 345 0 15

P
N U S
E
Mr
SUN
Me
V
J

Notes:
1) Planetary distances from the Sun are not in correct proportions, especially for the outer planets. Angles between planets are therefore only approximate.
2) Planets move in a counter-clockwise direction, spiraling out of the page, towards the observer, as the Solar System moves through the heavens.
3) True distances from the Sun, in astronomical units, and approximate revolutionary period.

Mercury	Me	0.38 AU	88 days to orbit once, or 3.7 days per 15 degrees
Venus	V	0.72 AU	225 days to orbit once, or 9.4 days per 15 degrees
Earth	E	1.00 AU	365 days to orbit once, or 15.2 days per 15 degrees
Mars	Mr	1.5 AU	687 days to orbit once, or 28.6 days per 15 degrees
Jupiter	J	5.2 AU	11.9 years to orbit once, or 181 days per 15 degrees
Saturn	S	9.5 AU	29.5 years to orbit once, or 1.2 years per 15 degrees
Uranus	U	19 AU	84 years to orbit once, or 3.5 years per 15 degrees
Neptune	N	30 AU	164.8 years to orbit once, or 6.9 years per 15 degrees
Pluto	P	39 AU	247.7 years to orbit once, or 10.3 years per 15 degrees

Research Progress Report

The following is a compilation of Research Progress Reports that were previously circulated to a small group of interested individuals and supporters, prior to the inception of the Pulse. Only a few minor changes and deletions have been made. From now on, all Research Progress Reports will be published in the Pulse.

RESEARCH PROGRESS REPORT #1, MAY 1987:

*   The Orgone Biophysical Research Laboratory, Inc. is a non-profit scientific research and educational foundation, established in 1978 by James DeMeo. DeMeo has acted as the primary Director of Laboratory activities since that date. In 1985, Robert Morris joined the Laboratory as Co-Director. Over the years, work undertaken by Laboratory staff has included research on the biophysical effects of the orgone energy accumulator, methods for objectification of the biological and atmospheric orgone energy, the effectiveness of the cloudbuster under both drought and non-drought conditions, and the effects of nuclear power plants and atomic explosions on weather and climate. Dr. DeMeo produced a graduate thesis on Reich's cloudbuster in 1979, while studying and teaching at the University of Kansas. This study documented various effects which had been noted previously by Reich and others. Later cloudbusting research undertaken in Kansas, Illinois, and Florida demonstrated the capacity of cloudbusting operations to terminate major droughts. The Laboratory is not formally affiliated with any other organization, but enjoys an informal, cooperative, and productive working relationship with other responsible researchers and organizations, such as the American College of Orgonomy, which has funded several Laboratory projects.

*   The Laboratory is developing a Drought Abatement Outreach Program, and a Desert Greening Program, to bring the techniques of cloudbusting to those areas where drought and aridity are at critical stages. In order to further the goals of these projects, plans are underway to develop a permanent field research station and laboratory facility in the arid zone of the Southwestern USA. Persons who wish to assist in this effort should contact either of the Laboratory directors.

*   This summer [1987], the Laboratory was generously given loan of a Meyers "Dark Invader" image intensifier. This intensifier will be experimentally used in a program to photograph both the small atmospheric orgone energy "units" or "particles", and the life energy fields of living and non-living materials. Photographs which are made will be digitized and image-enhanced via computers. Our thanks to Dr. Robert Haralick, professor of Electrical Engineering at the University of Washington, for making the loan, and for volunteering to participate in the image-enhancement phase of the experiment.

*   A bound xerographic copy of Dr. DeMeo's doctoral dissertation, titled "On the Origins and Diffusion of Patrism: the Saharasian Connection" has recently become available through University Microfilms International. Copies of this edition, as xerox facsimile or microfilm, can be purchased directly from University Microfilms (300 North Zeeb Road, Ann Arbor, MI 48106; telephone 800/521-0600) by giving the author's name, dissertation title, and the year (1986). This geographical work elucidates the origins of human armoring in the great desert belt of the old world at around 4000 BC. The dissertation will also be reprinted in serial form in forthcoming issues of the *Journal of Orgonomy.* [starting with the November 1987 issue, Volume 21, Number 2, and continuing]. This article series will be titled "Desertification and the Origins of Armoring: The Saharasian Connection". The work will also appear in a more popular book form sometime in the future.

RESEARCH PROGRESS REPORT #2, MAY 1988: DEVELOPMENT OF A DESERT RESEARCH FACILITY

Since January of 1988, the following progress has been made towards development of a Desert Research Center, previously outlined in a proposal circulated by James DeMeo, for the Orgone Biophysical Research Laboratory.

*   A 1988 Chevy Van has been purchased, which will be used as a primary tow vehicle for future cloudbusting excursions. The vehicle is equipped with heavy towing equipment, is air conditioned, and can comfortably sit four or five passengers, and carry a cargo.

*   A Leading Edge microcomputer (IBM Compatible) with 30 MB hard disk drive has been purchased, for use in data analysis, and for maintaining Laboratory records and mailing lists.

continued on page 46.

* The following progress has been made towards replication of the Earthquake/Atomic-Bomb-Test study of Kato:

1) A detailed data base on all Atomic Bomb tests has been obtained from the Swedish National Defense Research Institute, containing information on all above-ground and underground nuclear tests made by all nations since 1945, to include unannounced tests. Data includes time, date, location, and yield for each test.

2) The Laboratory has obtained access to the Earthquake Data Base System of the US Geological Survey, from which detailed global earthquake data is available. This data includes exact times, dates, locations and magnitudes for all earthquakes.

* Cloudbuster Icarus, constructed in 1976, was in need of major repairs to its frame, particularly following transportation from Miami to Iowa. These repairs have been completed; new tail lights and a jack wheel were also added to the unit, which continues to function well.

* A search for a remote desert facility of from 5 to 200 acres, for establishing the Desert Research Center, is presently underway. A preliminary field survey of the region has been made, and the search is on for a suitable site. Contacts are being developed with various individuals, governmental officials, non-profit groups involved in water resource issues, and agricultural cooperatives in the Southwest USA, to secure assistance and cooperation in the desert research.

* The Laboratory has acquired a number of research instruments necessary for the desert research work. These include wind speed and direction recording equipment, hydrothermographs, a microbarograph, precipitation gauges, a recording solar radiation meter, millivoltmeter, orgonotester, electroscope, 3" refracting telescope, and several strip chart recorders.

RESEARCH PROGRESS REPORT #3, OCTOBER 1988

* The Laboratory recently relocated to El Cerrito, California, on the eastern shore of San Francisco Bay. This move was prompted in part to an unfortunate failure to obtain adequate funding for the proposed Desert Research Center. This new location offers many advantages, however, as there are many people in the area who are familiar with Reich's works, and are supportive of other Laboratory projects and activities. Pending full funding of the proposed Center, the Laboratory will launch a series of short term experimental cloudbusting operations in the desert areas of the Western USA during 1989, with the goal of bringing moisture to those areas, and to prevent subsequent outbreaks of desert air into moister surrounding regions.

* A cloudbusting operation was performed on August 12-14 in the central core region of the American Southwestern desert, north of Yuma, Arizona. This operation was very successful in restoring flow of the galactic orgone stream (classically identified as the subtropical jet stream) into North America, stimulating rains across the desert Southwest, Southern Rocky Mountains, Great Plains and Midwest. A second cloudbusting operation was undertaken on September 11-15 in central Washington State. This operation restored the flow of moisture into the dry regions of the Pacific Northwest, and brought widespread rains to areas which had been suffering greatly from drought and forest fires. A full report on both of these operations will appear in a forthcoming issue [May 1989] of the *Journal of Orgonomy.*

* The Laboratory recently purchased a copy of the "Chicago Tornado Tape" data base compiled by Dr. T. Fujita of the University of Chicago. The tape, which contains data on all US tornadoes from 1900 to 1985, has been read to the Laboratory computer.

RESEARCH PROGRESS REPORT #4, MARCH 1989:

* Cloudbuster Icarus was in need of repairs following two major operations in Arizona and Washington last year. These repairs have been undertaken and completed, by Mr. Donald Bill. Thank you Don!

* The Laboratory will undertake a series of five experimental cloudbusting operations in the deserts of the American Southwest this summer. The purpose of these operations is to provide a better data base for the effects of desert cloudbusting, and as preliminary work towards a more permanent Desert Research Center, where an experimental Desert Greening Program can be undertaken. Funding for the basic expenses of these operations has been provided by the American College of Orgonomy. Thanks to Dr. Richard Blasband, and other members of the College, for arranging this needed support.

* The Laboratory is seeking funds for several specific research projects, as well as for general operating expenses, and equipment needs. A list of these can be provided to interested persons. We also are seeking donations of new or used bookshelves, filing cabinets, and other office equipment. Volunteers are also needed for various projects, such as filing, typing, doing library research, and for other matters. Individuals who can donate funds, or offer other tangible assistance, should contact Dr. James DeMeo or Theirrie Cook, at (415) 526-5978, or write to the Orgone Biophysical Research Laboratory, PO Box 1395, El Cerrito, CA 94530.

Cosmic Orgone Engineering Report

NOTE: The Orgone Biophysical Research Laboratory continues to receive requests from individuals on "how to" build and use a cloudbuster. Please note that the untrained and unskilled use of a cloudbuster can be dangerous to both the operator's health, and to the community in which it is used. We therefore cannot supply this kind of information. For a detailed overview of the kinds of problems that can occur from irresponsible use of a cloudbuster, and the kinds of research training and skills that are required for the safe and effective use of a cloudbuster, send $5 to the Laboratory for the booklet "So, You Want to Build a Cloudbuster?". This Laboratory, in cooperation with the American College of Orgonomy, and other responsible individuals and organizations, has established a Drought Abatement Outreach Program, to provide trained and skilled operators, and state-of-the-art cloudbusting equipment to those areas afflicted with drought. A Desert Greening Project is also in the planning stages. For further information, contact: The Orgone Biophysical Research Laboratory, PO Box 1395, El Cerrito, CA 94530.

PICTURE: Cloudbuster, "Icarus", on location.

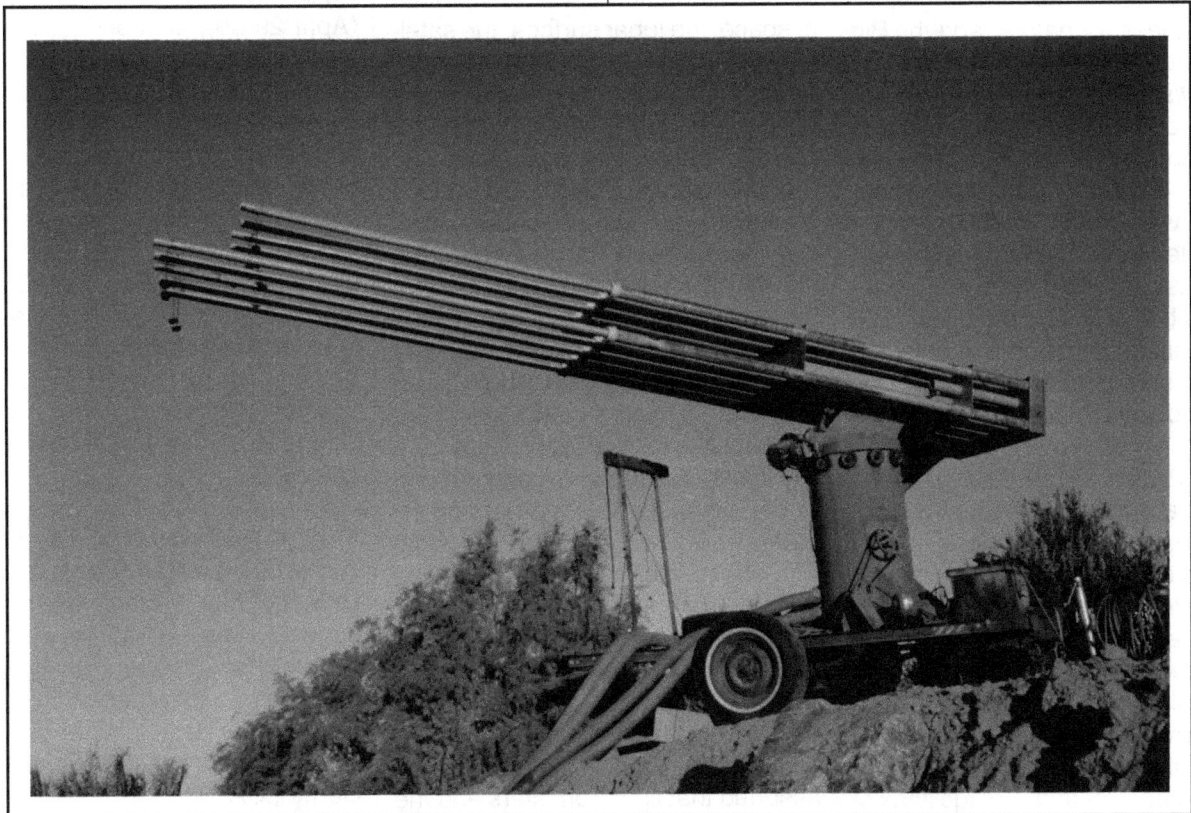

Science Notes

❖ Several recent astrophysical studies have confirmed the existence of unusual properties in supposedly "empty" space, all of which appear to have an energetic basis similar to Reich's orgone energy. Some aspects of these new findings suggest the existence of superimposing energy. For example, large scale "streaming motions" have been observed among galaxies (*Science*, 4 April 1986). The Earth has been calculated to be "drifting", in a net sense, through "radiation", at a rate of 600 km per second (Science, 20 June 1986). This velocity is nearly identical to the velocity of the "dynamic aether drift" previously discovered by Dayton Miller, and the directions of the "drifting" are along similar axes; a future article in the *Pulse* will discuss Miller's work. Huge and strange "luminous arcs", larger than galaxies (according to conventional distance-size determinations) have also been photographed in deep space, in violation of theories on the thermodynamic "randomness" of nature, and the Big Bang theory (*Science*, 6 February 1987; *Discover*, June 1988). Photographs of unusual, superabundant "blue spots" have also been made in the deepest depths of space, where no otherwise clearly identifiable stars or galaxies exist. Called "blue fuzzies" by the astronomers, they are postulated as being "young galaxies". But again, this explanation is predicated upon use of the conventional distance indicators, and also upon the assumption of "empty space" (Science, 22 July 1988). All of these findings appear to be expressions of the orgone energy in space, or the corpuscular *orgone units* in the atmosphere (which would also impinge upon the film of the astronomers, inside their telescope cameras) as previously described by Reich. To date, however, we do not see any honest discussion of the idea that a dynamic energy might actually exist in space, to account for these observations. All the theories that are being postulated, such as "cosmic strings" and "big foams", are bending over backward in an effort to preserve the emptiness of space.

❖ An interesting question has recently been raised, which will be put to our readers. It seems that stars might not be recorded on photographic film or videotape when the camera is taken into space, outside of the Earth's atmosphere and orgone envelope. When we first heard of this idea, it sounded hard to believe... until we searched through photographic records dating back to the Apollo program. We reviewed many books, compendiums, and slide sets of published NASA photos, taken from space and on the surface of the moon by hand-held cameras used by the astronauts, and we have not been able to find a single star in any of them. One can see the pitch-black background of space, the lunar surface, the satellites and other equipment of the astronauts, but *no stars, anywhere.* This is so, even in photos where the shutter speed of the camera would have had a slower setting. From the various quotations by the astronauts, it is clear that they could *see* stars, but none of the quotations suggest that the astronauts were awed by looking at the stars or the galaxy, that they were exceptionally beautiful or grand to look at, as is the case when seen from Earth on a clear night, away from city lights. Readers are encouraged to double-check this matter, and let us know if they find a space photo with stars. Warning: don't be fooled by "star-like" debris floating near recently-separated satellites or booster rockets, or by dust on the photos (which can be brushed away). Perhaps there is a straightforward explanation for this, but we note that Reich asserted that light from stars and the sun was a local phenomena, resultant from the process of excitation and lumination of the Earth's orgone energy envelope. If this observation holds, and there is no clear evidence of stars being photographed in space, it would vindicate Reich's hypothesis in a powerful and unexpected manner.

Announcements:

❖ The **Reich Blood Test** for the estimation of bioenergetic charge is available by appointment at the Elsworth F. Baker Oranur Research Laboratory in Princeton, NJ. For more information, contact Richard A. Blasband, M.D., American College of Orgonomy, PO Box 490, Princeton, NJ 08542.

❖ A series of introductory weekend workshops, focused on "**The Bioenergetic, Orgonomic Basis of Life and Weather**", led by James DeMeo, Ph.D., are scheduled for Boulder, CO (April 29-30), Cambridge, MA (May 20-21), Seattle, WA (June 17-18), Atlanta, GA (July 15-16), and Berkeley, CA (Aug. 26-27). They will cover both sex-economic and orgone biophysical subject materials, reviewing the works of Reich, and the more recent confirming and extending research of others. Fee is $190, or $90 for full-time students. For a detailed flyer, and registration materials, contact Dr. James DeMeo, Orgone Biophysical Research Laboratory, PO Box 1395, El Cerrito, CA 94530, (415) 526-5978.

❖ A new edition of *The Orgone Accumulator Handbook*, by James DeMeo, Ph.D., will be published after June 15. The new edition will feature several new chapters and appendices, plus an introduction by Dr. Eva Reich. For details, write to Natural Energy Works, ~~PO Box 1395, El Cerrito, CA 94530~~.

Orgonomic Research Review, 1988 to Present

Appearing in the Journal of Orgonomy 22(1), May 1988:

- Further Problems of Work Democracy (Part II), by Wilhelm Reich, M.D.
- In Seminar with Dr. Elsworth Baker
- CORE Progress Report #17, Fighting Forest Fires and Breaking the 1987 Drought in the Northwest U.S., by John Schleining, M.S.
- Armoring in Women in Labor: A Perinatal Research Group Report, by Richard A. Blasband, M.D., Charles Konia, M.D., and Robin R. Karpf, M.D.
- The Orgone Energy Light: A Pilot Experiment, by Richard A. Blasband, M.D.
- Culturing SAPA Bions, by Kevin Carey, B.S. and Steven Dunlap
- Orgone Therapy (Part VI), by Charles Konia, M.D.
- A Case of Spastic Dysphonia, by Richard Schwartzman, D.O.
- Orgonomic First Aid in the Treatment of Neurologic Disaster, by Howard J. Chavis, M.D.
- Desertification and the Origins of Armoring: The Saharasian Connection (Part II), by James DeMeo, Ph.D.
- Transformations in Microbiological Organisms, by Richard A. Blasband, M.D.

Appearing in the *Journal of Orgonomy*, 22(2), November 1988:

- Further Problems of Work Democracy (Part III), by Wilhelm Reich, M.D.
- In Seminar with Dr. Elsworth Baker
- The Emotional Plague as Manifested in the AIDS Hysteria, by Robert A. Harman, M.D.
- An Evaluation of the Risk of AIDS Transmission, by Robert A. Harman, M.D.
- Bionous Tissue Disintegration in Three Patients with AIDS, by Alan R. Cantwell, Jr., M.D. and Richard A. Blasband, M.D.
- The Creation of Matter in Galaxies, by Charles Konia, M.D.
- Genitality Achieved by a Passive Feminine, by Charles Konia, M.D.
- Orgonomic First Aid for Eating Disturbances in Medical Illness, by Howard J. Chavis, M.D.
- Orgone Therapy (Part VII), by Charles Konia, M.D.
- Desertification and the Origins of Armoring: The Saharasian Connection (Part III), by James DeMeo, Ph.D.
- CORE Progress Report #18: Fighting the Great Drought of the Summer of 1988, by Richard A. Blasband, M.D.

- For the Record: Transformations in Microbiological Organisms, by Richard A. Blasband, M.D.

Appearing in *Annals, Institute for Orgonomic Science;* (5):1, September 1988:

- Reich's Bioelectric Experiments: A Review With Recent Data, by Byron Braid, M.D. & Robert Dew, M.D.
- Human Armoring: An Introduction to Psychiatric Orgone Therapy, by Morton Herskowitz, D.O.
- The Management of a Case of Substitute Contact, by Michael Ganz, M.D.
- A Case of Longstanding Borderline to Mild Hypertension, by Arthur Nelson, M.D.
- Considerations in the Treatment of Ocular Armoring, by David Schwendeman, M.D.

Appearing in Other Publications:

- Northwest Drought and the North Idaho Crop: August to October 1988, by George D. Conrad, B.A.,J.D., *Association for Life Education J.,* 2(1): 3-11, January 1989.
- Progress on Caesareans?, by Patricia M. Coe, M.A., *Association for Life Education J.,* 2(1):13-15, January 1989.
- Cloudbusting, Fact or Mythology, by James DeMeo, Ph.D., in *Acres, USA,* August 1988, p.1, and *Wildfire,* 4(1):12-14, Fall, 1988.
- Interview with James DeMeo, *Wildfire,* 4(2):14-27, Winter, 1988.
- Effects of Orgonotic Devices on Tomato Plant Growth, by A. Parimal Sellers, *Living Tree Journal,* 2:53-61, 1988-1989.

Orgonomic Publications

*   *Journal of Orgonomy*, PO Box 490, Princeton, NJ 08524

*   *Annals, Institute for Orgonomic Science,* PO Box 304, Gwynedd Valley, PA 19437

*   *Offshoots of Orgonomy,* PO Box 1248, Gracie Station, New York, NY 10028

*   *Association for Life Education Journal,* PO box 383, Careywood, ID 83809

*   Wilhelm Reich Museum and Bookstore, PO Box 687, Rangely, ME 04970

Calendar of Forthcoming Events, with Contact Information:

1989

April 8-9, April 12-13, April 14: **West Berlin, Germany,** Weekend workshops and public lecture, Dr. James DeMeo, in English, Reich Society, contact Dr. Heiko Lassek, Germany telephone 030-891-4914.

April 15-16: **Princeton, New Jersey**, Introductory Laboratory Workshop on Orgone Biophysics, Teaching staff: Dr's. Crist, Harman, Karpf, Konia, Schleining, American College of Orgonomy, PO Box 490, Princeton, NJ 08542, (201) 821-1144.

April 20-22: **Amsterdam, The Netherlands,** Congress on Geo-Cosmic Relations, various speakers, including "The Discovery of a Cosmic- Atmospheric-Biological Energy Principle", by James DeMeo, Ph.D., contact Dr. G. Tomassen, Netherlands telephone 31.83.70.8.35.22.

April 29-30: **Boulder, Colorado**, Introductory weekend workshop, "The Bioenergetic, Orgonomic Basis of Life and Weather", Dr. James DeMeo, Orgone Biophysical Research Laboratory, PO Box 1395, El Cerrito, CA 94530, (415) 526-5978.

May 20-21: **Cambridge, Massachusetts**, Introductory weekend workshop, "The Bioenergetic, Orgonomic Basis of Life and Weather", Dr. James DeMeo, Orgone Biophysical Research Laboratory, PO Box 1395, El Cerrito, CA 94530, (415) 526-5978, or Dr. Myron Sharaf (617) 969-1653.

June 3-6: **Princeton, New Jersey**, Advanced Laboratory Workshop in Orgone Biophysics, Teaching staff: Dr's. Crist, Harman, Karpf, Konia, Schleining, American College of Orgonomy, PO Box 490, Princeton, NJ 08542, (201) 821-1144.

June 17-18: **Seattle, Washington**, Introductory weekend workshop, "The Bioenergetic, Orgonomic Basis of Life and Weather", Dr. James DeMeo, Orgone Biophysical Research Laboratory, PO Box 1395, El Cerrito, CA 94530, (415) 526-5978.

July 15-16: **Atlanta, Georgia,** Introductory weekend workshop, "The Bioenergetic, Orgonomic Basis of Life and Weather", Dr. James DeMeo, Orgone Biophysical Research Laboratory, PO Box 1395, El Cerrito, CA 94530, (415) 526-5978.

July 28-30: **Los Angeles, California,** Annual Meeting of the Cancer Control Society; Dozens of speakers on the topics of non-toxic, alternative treatments for degenerative disease. A lecture on the "Discovery of the Life Energy" will be given by Dr. James DeMeo. Contact Lorraine Rosenthal of the Cancer Control Society for details, (213) 663-7801.

August 26-27: **Berkeley, California,** Introductory weekend workshop, "The Bioenergetic, Orgonomic Basis of Life and Weather", Dr. James DeMeo, Orgone Biophysical Research Laboratory, PO Box 1395, El Cerrito, CA 94530, (415) 526-5978.

October: **Tokyo, Kansai, Kanazawa, Japan,** Introductory weekend workshops, Dr. James DeMeo, Center for Bioenergy, Kanazawa, contact Dr. Dennis Hoerner, Japan telephone (0762) 24-4892, or the Orgone Biophysical Research Laboratory, PO Box 1395, El Cerrito, CA 94530, (415) 526-5978.

1990:

June 14-17: **Nice, France,** Fifth International Orgonomic Conference, various speakers and topics, contact Dr. Giuseppe Cammarella, telephone 93.81.96.96, or c/o American College of Orgonomy, PO Box 490, Princeton, NJ 08542, USA , telephone (201) 821-1144.

Letters to the Editor

To the Editor,

I have long wondered about Reich's hypothesis on the Oranur Experiment's effects around Orgonon and much of the northeast in January and February of 1951. In particular, he noted the detonation of several atomic bombs in Nevada during the period from 27 January to 6 February, and argued that these may not have been responsible for a high background radiation count for hundreds of miles around Orgonon from 29 January to 3 February. He reasoned as follows:

> "We had felt responsible for possible chain reactions in the atmospheric energy around Orgonon long before the atomic blasts occurred... If we assumed that the higher background count in the eastern U.S.A. was *not* due to oranur, but was caused by the atomic explosions in Nevada, the following inconsistencies existed:
>
> a)...the high count had already been observed for several days *before* Feb.3, 1951, i.e., only two to three day;s after the first explosion.
>
> b) The increased radioactivity in the atmosphere at Rochester, New York was found in snow that had fallen...thus, radioactivity had supposedly travelled the 2300 miles (!!) from the Las Vegas area in Nevada to the east in only two or three days, or with a speed of about 1200 miles a day, fifty miles per hour, i.e. with the speed of a whirling hurricane, on clear, windless days, faster than an average hurricane, which progresses at a rate of only ten to twelve miles an hour...
>
> c) The increase of atmospheric radioactivity had been noticed only in the east. From Rochester, New York, to Las Vegas, Nevada, with the exception of the immediate vicinity of Las Vegas, nothing unusual had been noticed. Is it possible that the radioactive "cloud" traveled with the speed of a major storm over 2300 miles, leaving no trace until it reached the Eastern border states...? I believe such an interpretation is far less acceptable than the other one — that oranur was responsible for the increased atmospheric activity.
>
> d) Most reports available so far on atomic explosions stress the fact that the high radioactivity lasts only a few seconds, that it reaches only a few miles beyond point zero. I have heard of no effect occurring as far as 2300 miles away, *with*

an untouched area of some 1700 miles between the blast and the location of increased radioactivity. On the other hand, reports from Bikini state that living organisms remained highly radioactive for years after the explosion there..." (Reich, W.: *Selected Writings*, Farrar, Straus & Giroux, pp.387-388, 1975.)

Based on some new evidence, only recently available to the public, I have come to conclude that Reich was mistaken on this point, and in order to clarify our attempts to understand oranur, I though it would be useful to share this with your readers who may not yet have seen it. The evidence can be found in a 1986 work on atmospheric nuclear tests called *Under the Cloud,* by Richard L. Miller (Macmillan Free Press, NY). Among the most valuable data in this study are maps of the paths taken by the fallout clouds resulting from each of the first 70+ nuclear detonations in Nevada, and in many cases also detailed descriptions of the height into the stratosphere to which fallout was carried and the speed of winds carrying the clouds of radioactive debris. It seems that the military was interested enough in following the path of the fallout to have planes aloft with instruments to track the clouds and measure their speed and direction of travel. The facts which were unavailable to Reich (and the rest of the American public) included that winds speeds at such high altitudes (in some cases the radioactive debris was carried more than 50,000 feet up) were very great — up to 200 mph or more. The cloud from the first Nevada test, on 27 January, did not get so high or travel quite this fast; nonetheless, tracking planes had followed it over Colorado only 12 hours after the blast, and across Kansas and Missouri at the same speed.

This cloud continued moving eastward (Miller, pp.89-92), reaching the area of New York and Massachusetts "sometime after 9:00 AM Eastern Standard Time on January 29" at which time it was "at least 300 miles wide". This cloud of debris clearly *could* have been a mechanical cause of the high radioactivity readings Reich read about, and believed were due to oranur. It is also possible that the presence of the radioactive debris from this and successive tests, which also arrived over the next few days, could have interacted in some functional way with Reich's experiment, though the description of the tests done to characterize the fallout in Rochester, New York seems to argue more persuasively for the former explanation (Miller, p.90). At the very least, we must revise Reich's mistaken view that the fallout itself could not have arrived so quickly, and take into account the knowledge

Continued on Page 52.

that it was physically present over New England by 29 January, in our future attempts to understand the results of the Oranur Experiment. I highly recommend Miller's book as a wealth of data for anyone trying to comprehend the degenerating effects of nuclear testing on the atmosphere.

Reich's point d) from the quote above is instructive in another light. It illustrates the falsity of the information which the Atomic Energy Commission was making available to the public in the late 1940s and early 1950s. Though their scientists were studying fallout, the A.E.C. either underestimated its dangers, or were deliberately concealing the dangers from the public. Reich seems to have been one of the first scientists to recognize that nuclear testing might have much more serious consequences than were being publicly acknowledged. This is a tribute to his astuteness as an *observer*, notwithstanding the possible incorrectness of his specific hypothesis, since he, like most scientists, was being denied the data necessary to test such hypotheses. This alarming state of overall ignorance in atmospheric problems in the scientific community persisted for many years and was never more clearly exemplified than at the 1983 conference on the *Nuclear Winter*. A questioner in the audience asked "why the climatic consequences (of nuclear weapons explosions) had not been discovered before". The panel, including many of the most prestigious atmospheric scientists in the world, replied that *they couldn't answer the question,* though noting that "the basic physics and chemistry had been available for 20 years, and governments had a responsibility to study them". (*Bulletin of Atomic Scientists*, April 1984, p.85)

Another recent work, *Radiant Science, Dark Politics,* by Martin Kamen, sheds light on another chapter of Reich's work which has been puzzling to me: the manner in which Einstein so abruptly broke off contact with Reich. Kamen's account of the irrational suspicion which focused on him and any scientists even remotely connected with the Manhattan Project, even if they only had acquaintances on the political left, confirms Jerome Greenfield's hypothesis (J. Orgonomy, 16:93) that Einstein was probably instructed to break off contact with Reich. Such an episode would be entirely "in character" for the FBI, as seen in Kamen's case, or as also seen in the case of Oppenheimer (e.g. R. Jungk, *Brighter than 1000 Suns,* Harcourt Brace Jovanovich, NY).

I would be interested in any comments, and hope these thoughts will prove useful to others.

James Strick
Science Department
Park School
Baltimore, Maryland

Editor's Response:
[When radioactivity from the Chernobyl nuclear disaster arrived in the USA, carried by stratospheric winds, reports of measured radioactivity at the ground surface were very irregular. The same was true regarding radiation from Chinese atmospheric bomb tests in the 1970s. Some states reported measured radiation, while other nearby states did not. But was the absence of reported radioactivity in some states based upon an absence of measured high readings, or a *failure to report* measured high readings? Or, perhaps no measurements were taken at all!? We may never know for sure. Regarding Reich's observations, what is needed most is more research into methods (including Reich's OR-charged GM techniques) to measure the physically sensible effects of oranur from nuclear power plants and atomic explosions, to better differentiate oranur from the mechanical effects of "radioactive debris" - J.D.]

In Remembrance

Jerome Eden

(1925 - 1989)

Orgonomy lost a dedicated and courageous advocate when Jerome Eden died on January 18th, after a long battle with scleroderma and cancer. He was sixty-three. Mr. Eden's contributions to orgonomy over the years were many. First, as an author, he wrote ten books related to orgonomy... Perhaps the best known of these, *Orgone Energy, The Answer to Atomic Suicide,* was published in 1974 as an introduction to Dr. Reich's work and the dangers of nuclear energy. In addition to his books, Mr. Eden published several journals and contributed many articles to the Journal of Orgonomy.

Mr. Eden also was an expert cloudbuster. Using "Bluebird", he conducted numerous cloudbusting operations from his home in Careywood, Idaho. This past summer, though too ill himself to operate the cloudbuster, he directed a team of his students in a successful effort to alleviate the drought in the Inland Northwest.

Perhaps more than anything else, however, Mr. Eden was a pioneer in the field of orgonomic UFOlogy. Virtually alone among orgonomists, he took Dr. Reich's UFO work seriously and followed it up with his own research. Some of his insights are contained in two of his books *Planet in Trouble - The UFO Assault on Earth*, and *Scavengers From Space*.

Jerome Eden was a fighter for life. He battled the Atomic Energy Commission, took on life-negating public schools, and fought the emotional plague within orgonomy and without. Orgonomy will miss him, and his unswerving dedication to life.

Jonathan S. Coe
Careywood, Idaho

[Editor's note: Mr. Eden attributed his illness to the effects of overcharge and DOR radiation from years of working with a manually-operated cloudbuster. It is notable that in the last months of his life, he continued working, and was very much occupied with the drought and forest fire situation in the Pacific Northwest. Books by Jerome Eden may be ordered directly from the *Association for Life Education*, PO Box 383, Careywood, ID 83809.]

✢ *The Orgone Accumulator Handbook,* by James DeMeo, Ph.D., excerpts of a chapter from a new book.

✢ *Dangers of Infant Circumcision*, a chapter from a new book by Marilyn Milos, Director of the National Organization of Circumcision Information Resource Centers (NOCIRC).

✢ *The Ether-Drift Experiment and the Determination of the Absolute Motion of the Earth*, by Dayton Miller; reprint of an important study.

✢ A comprehensive review of National Weather Service rainfall data collected during past cloudbusting operations, showing the restoration of atmospheric pulsation and rains.

✢ Evidence of increases in rainfall from cloudbusting operations in the Negev Desert, Israel.

✢ A report on several conferences devoted to new research discoveries where Reich's work was enthusiastically received and discussed.

✢ A review of Gorgio Piccardi's independent discovery of the orgone accumulator principle.

✢ A review of Louis Kervran's findings on the transmutation of stable, non-radioactive elements by living creatures.

✢ A discussion of new methods for measuring the biological and atmospheric orgone energy.

✢ More on the effects of underground atomic explosions on weather and people.

✢ Plus all our regular features, book reviews, reports, and more...

FOR ADDITIONAL INFORMATION AND BOOKS, by Wilhelm Reich and others on the subjects of sex-economy, bioelectricity, life energy and orgone energy, plus relevant research instruments, contact:

NATURAL ENERGY WORKS,
~~PO Box 1395,~~
~~El Cerrito, CA 94530~~
~~(request a copy of their catalog ($1).~~

PULSE OF THE PLANET is looking for graphics artists and cartoonists to donate small line drawings and other art for possible inclusion. All materials that are used will be fully acknowledged.

NOTE: For an update on Dr. DeMeo's research since this early 1989 publication, please review these websites:

* James DeMeo's Research Publications:
    https://orgonelab.academia.edu/JamesDeMeo

* Orgone Biophysical Research Lab:
    http://www.orgonelab.org

* Saharasia web page:
    http://www.saharasia.org

* Natural Energy Works, books:
    http://www.naturalenergyworks.net

www.ingramcontent.com/pod-product-compliance
Lightning Source LLC
Chambersburg PA
CBHW080056280326

41934CB00014B/3334